Chaotic Modelling and Simulation

Analysis of Chaotic Models, Attractors and Forms

Christos H. Skiadas

Charilaos Skiadas

CRC Press
Taylor & Francis Group
Boca Raton London New York

CRC Press is an imprint of the
Taylor & Francis Group, an **informa** business

A CHAPMAN & HALL BOOK

CRC Press
Taylor & Francis Group
6000 Broken Sound Parkway NW, Suite 300
Boca Raton, FL 33487-2742

First issued in paperback 2019

© 2009 by Taylor & Francis Group, LLC
CRC Press is an imprint of Taylor & Francis Group, an Informa business

No claim to original U.S. Government works

ISBN-13: 978-0-4200-7900-5 (hbk)
ISBN-13: 978-0-367-38665-8 (pbk)

Library of Congress Cataloging-in-Publication Data

Skiadas, Christos H.
     Chaotic modelling and simulation: analysis of chaotic models, attractors and forms / Christos H. Skiadas, Charilaos Skiadas.
         p. cm.
     Includes bibliographical references and index.
     ISBN 978-1-4200-7900-5 (hardcover : alk. paper)
     1. Chaotic behavior in systems. 2. Mathematical models. I. Skiadas, Charilaos. II. Title.

Q172.5.C45S53 2009
003'.857--dc22   2008038183

Visit the Taylor & Francis Web site at
http://www.taylorandfrancis.com

and the CRC Press Web site at
http://www.crcpress.com

Preface

"Is there a chaotic world? To what extent? Mathematics, geometry or simply chaos?"

The above questions arise time and time again when dealing with chaotic phenomena and the related attractors and chaotic forms. During the lengthy preparation of this manuscript, computers were used in the same way that geometers of old used geometric tools. Simple or complicated ideas are designed, estimated and tested using this new tool. The result is more than 500 graphs and illustrations that fill the pages in front of you.

Verbal explanations are kept to a minimum, and a lot of effort is devoted to ensuring that the presentation of ideas progresses from the most elementary to the most advanced, in a clear and comprehensible way, while requiring little previous knowledge of mathematics. In this way, we hope that a more general audience will benefit from the material of this book, especially from the large variety of chaotic attractors included.

A lot of new material is presented alongside the classical forms and attractors appearing on the literature on chaos. However, most of the illustrations are based on new simulation methods and techniques. A rather lengthy introduction provides the reader with an overview of the subsequent chapters, and the interconnection between them.

Our inspirations came from various disciplines, including geometry from the ancient Greek and Alexandrian period (a great deal of which is taught in the Greek school system), mathematics from the developments of the last centuries, astronomy and astrophysics from recent developments (the Conference on Chaos in Astronomy organized by George Contopoulos in 2002 was very stimulating) and the amazing chaotic illustrations of Hubble and Chandra and the vortex movements from fluid flow theory.

Chaotic theory is developing in a new way that influences the world around us, and consequently also influences our ways of approaching, analyzing and solving problems. It is not surprising that one of the central models in the chaos literature, the Hénon-Heiles model, is presented in a paper with the title *"The applicability of the third integral of motion: Some numerical experiments."* Numerical experiments in 1964 were the basis for many significant changes in astronomy in the decades that followed. In 1963 Edwin Lorenz, in his pioneering work on *"Deterministic Nonperiodic Flow,"* proposed a more prominent title for chaotic modelling, by including the term "deterministic." His work spearheaded numerous studies on chaotic phenomena. Fifteen years later, Feigenbaum explored the rich chaotic properties of the simple discrete Logistic model, known since 1845 following the work of Verhulst on the continuous version of this model, and introduced scientists to the chaotic field.

Our intention in this work was to present the main models developed by the pioneers of chaos theory, along with new extensions and variations to these models. The plethora of new models that can arise from the existing ones through analysis and simulation was surprising.

This book is suitable for a wide range of readers, including systems analysts, mathematicians, astronomers, engineers and people in any field of science and technology whose work involves modelling of systems. It is our hope that this book will prompt more people to become involved in the rapidly advancing field of chaotic models and will inspire new developments in the field.

Christos H. Skiadas
Charilaos Skiadas

List of Figures

Contents

Chapter 1

Introduction

1.1 Chaos in Differential Equations Systems

Four centuries ago, Newton and other scientists introduced the idea of *determinism* in the mathematical representation of the real world. As a result, future events in nature could be explained using knowledge of the past. The movements of stellar bodies were assumed to follow predetermined paths according to laws that were known or had to be discovered by the scientific community. The theory of determinism also had theoretical support from philosophy and other scientific fields. It would be difficult to say that determinism, as a general idea, can solve all problems in nature. The scientists of the last few centuries followed this theory in order to find rational routes in the construction of models. It was essential to simplify our description of the world first, and only then to solve fundamental problems, instead of entering into obscure paths of uncertainty and later on Chaos.

How then does one introduce a theory that goes against the perfect world of Newtonian dynamics? Our solar system has been rotating for millions of years, and the experience of the last centuries was in favor of determinism. The sun continues to rise every day, the moon follows a four-week period, the stars of Ursa Major are found every year in predetermined places. How and why could this perfect deterministic world lose this certainty and lead a chaotic existence?

It was not easy to convince the scientific community to introduce uncertainty into this deterministic world and accept chaotic theories. However, advances in applied mathematics made this transition easier. At the end of the nineteenth century, the *Runge-Kutta* method for numerical solutions to differential equations dramatically improved our ability to get more accurate solutions for numerous models. It was a real milestone in applied mathematics, especially for the study of highly non-linear problems, for which the earlier method by Euler was not very accurate.

Following the introduction of the Runge-Kutta method, the paths traversed by a stellar body could now be estimated quite efficiently. Although this method was still quite laborious at those early days, it gave new directions in the way scientists addressed various problems. Theory and practice could now equally benefit each other.

It is no surprise that this change influenced even the French school, traditionally the most theoretical of all, especially in mathematics and physics. Henri Poincaré, one of the most prominent scientists in 19th century France, tried to resolve the famous three-body problem in a contest proposed by the King of Norway (Poincaré, 1890b, 1892). Although he succeeded in solving a special case of this problem, namely the movement of a small mass around two larger masses revolving in two dimensions, he noticed that the orbits of three bodies moving under a central force due to gravity are quite complicated and can change drastically under small changes of the initial conditions.

Nowadays we know that to solve this problem we need a set of more than three coupled differential equations, and the paths of the solutions have a chaotic character. Poincaré tried to determine solutions to the restricted three-body problem, when one mass is very small relative to the other two. He understood that, even in this simple case, the motion had still a very complicated form. He tried to explain the phenomenon more generally and to establish a theory describing the chaotic paths of the solutions to a system of differential equations: The three-dimensional solutions to a system of non-linear differential equations could be chaotic. In stark contrast, the two-dimensional paths do not exhibit a chaotic behaviour, as follows from the famous Poincaré-Bendixson theorem: Chaos cannot occur in a two-dimensional autonomous system.

To elaborate a bit, consider the two-dimensional autonomous system (1.1):

$$\dot{x} = f(x, y)$$
$$\dot{y} = g(x, y)$$

(1.1)

where $\dot{x}$ and $\dot{y}$ are the first time derivatives. The system (1.1) can take the form of a two-dimensional equation for x and y:

$$\frac{dy}{dx} = \frac{g(x, y)}{f(x, y)}$$

(1.2)

and we can find phase paths in the plane describing the motion, directly from 1.2. Assuming that we have determined the *equilibrium points* at which both $f(x, y) = 0$ and $g(x, y) = 0$, the Poincaré-Bendixson theorem says that a phase path will either:

a) terminate at an equilibrium point,

b) return to the original point, resulting in a closed path, or

c) approach a limit cycle.

Hence, there are no chaotic solutions. The system of differential equations (1.3) illustrates all three cases:

$$\dot{x} = -y(y - a)(y - b)$$
$$\dot{y} = x(x - a)(x - b)$$

(1.3)

The resulting differential equation for x, y is:

$$\frac{dy}{dx} = -\frac{x(x - a)(x - b)}{y(y - a)(y - b)}$$

This equation can be easily integrated, resulting in the solution:

$$\frac{x^4 + y^4}{4} - (a+b)\frac{x^3 + y^3}{3} + ab\frac{x^2 + y^2}{2} = h$$

where h is the constant of integration. The paths, with parameters equal to $a = 0.8$ and $b = 0.3$, are illustrated in Figure 1.1. There are nine equilibrium points, four at centres four corners, four in the sides, and one in the middle of the figure. The more complex paths pass from the four unstable equilibrium points at the sides. Such systems, not chaotic in themselves but nonetheless not without interest, are further discussed in Chapter 7.

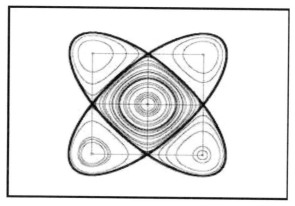

FIGURE 1.1:   The (x, y) diagram

On the other hand, the simplest three-dimensional non-linear system of differential equations, the Rössler model (Rössler, 1976d)

$$
\begin{aligned}
\dot{x} &= -y - z \\
\dot{y} &= x + ey \\
\dot{z} &= f + xz - mz
\end{aligned}
\tag{1.4}
$$

easily produces chaotic paths for appropriate values of the parameters.

The Rössler system has only one non-linear term, in the last equation. However, when the parameters have values $e = f = 0.2$ and $m = 5.7$, chaotic paths appear with the form presented in Figure 1.2(a). The Rössler system, along with other interesting three-dimensional systems, is examined in greater detail in Chapter 6.

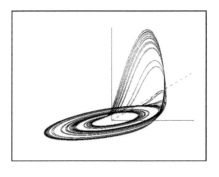
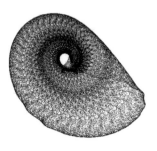

(a) The continuous case (x, y) (b) A discrete analogue

FIGURE 1.2:   Continuous and discrete Rössler systems

1.2   Chaos in Difference Equation Systems

Radical changes in the behaviour appear when a system of differential equations is replaced by a similar system of difference equations, or by a single difference equation with non-linear terms. For example, suppose we replace the Rössler system (1.4) by a corresponding system of difference equations:

$$x_{t+1} = x_t - \frac{1}{6}(y_t + z_t)$$
$$y_{t+1} = y_t + \frac{1}{6}(x_{t+1} + ey_t) \tag{1.5}$$
$$z_{t+1} = z_t + 0.1 + ex_{t+1}z_t$$

Then, for parameter value $e = 0.2$, and iterating under the condition $x_{t+1} = x_t - 0.4$, an interesting chaotic form like a sea-shell results (Figure 1.2(b)).

For centuries the mathematical development of methods and tools of analysis concentrated mainly on differential equations, and not on difference equations. It is therefore not surprising that the *logistic model*, the most popular chaotic difference equation model today, was "invented" only in the last three decades. The logistic model is described by a very simple equation:

$$x_{t+1} = bx_t(1 - x_t) \tag{1.6}$$

This very simple model already exhibits the key elements of chaotic behaviour. As the parameter b increases, period doubling bifurcations occur, giving rise to stable orbits of ever increasing size, culminating in completely chaotic behaviour. On the other hand, the differential equation analogue of this model was very popular:

$$\dot{x} = bx(1 - x) \tag{1.7}$$

Equation (1.7), as expected, gives simple finite solutions without complications and chaotic behaviour. This continuous model, called the *logistic model*, was applied in demography by Verhulst already in 1838.

Chapter 2, further discusses models and the modelling process in general. Deterministic, stochastic and chaotic models are introduced, along with chaotic analysis and simulation techniques. The subsequent chapters deal with particular classes of models, and their interesting behaviour.

1.2.1   The logistic map

Chapter 3 is devoted to a detailed analysis and simulation of the chaotic behaviour of non-linear difference equations, starting with the *logistic* model. The logistic map and its bifurcation diagram are presented. The bifurcation diagram exhibits the stable orbit of the x values for each value of the chaotic parameter b. In the case of the logistic model, the stable orbits become more and more complex as b increases, culminating in the complete chaos observed for b close to 4.

Other models that are examined, and exhibit similar behaviour, are: the *sinusoidal* model, the *Gompertz* model and a model proposed by *May*. Also, a class of models with behaviour different than that of the quadratic maps are studied, the *Gaussian-type* models. As can be seen clearly from the bifurcation diagrams of the logistic (Figure 1.3(a)) and the Gaussian model (Figure 1.3(b)), there are clear differences between the two classes of models, as well as some similarities in the chaotic windows.

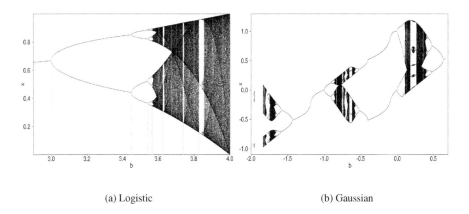

(a) Logistic   (b) Gaussian

FIGURE 1.3:   Bifurcations of the logistic and the Gaussian

1.2.2   Delay models

Delay models are models where future values depend on values far in the past. For instance, a simple delay equation is the *delay logistic*:

$$x_{t+1} = bx_t(1 - x_{t-m})$$

which is analysed further in Chapter 4, along with some variations and more complicated models.

Delay models can be found in several real life cases when the future development of a phenomenon is influenced from the past. The complexity of the delay system is higher when the delay is large and the equations used are highly non-linear. Such chaotic behaviour is expressed by a model proposed by Glass

$$x_{t+1} = cx_t + a\frac{x_{t-m}}{1 + x_{t-m}^{10}}$$

When the parameters are $a = 0.2$, $c = 0.9$ and the delay $m = 30$, the Glass model gives a very complicated graph in the phase space and chaotic time oscillations of a very complex character (Figure 1.4(a)). Figure 1.4(b) presents the map of the delay logistic, as well as the related oscillations, when the delay is $m = 10$. The chaotic parameter is $b = 1.2865$.

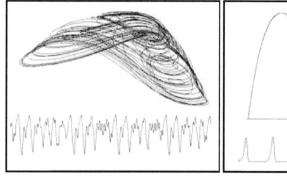
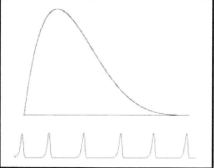

(a) The Glass model   (b) The delay logistic

FIGURE 1.4:   Delay models

1.2.3   The Hénon model

In some cases a system of two difference equations is replaced by a single difference delay equation with the same properties. This is the case of the Hénon model (Hénon, 1976), which can be thought of either as a system of two difference equations, or as a single delay equation. The model proposed by Hénon, which is now considered one of the standard models in the chaos literature, has the form:

$$x_{t+1} = 1 - ax_t^2 + y_t$$
$$y_{t+1} = bx_t$$
(1.8)

The equivalent delay difference equation of the Hénon system is:

$$x_{t+1} = bx_{t-1} + 1 - ax_t^2$$
(1.9)

For parameter values $a = 1.4$ and $b = 0.3$ the famous Hénon attractor is formed (Figure 1.5(a)).

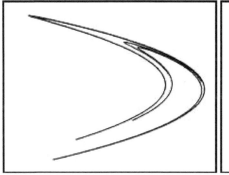 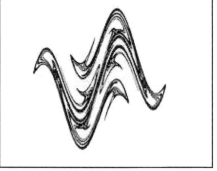

(a) The Hénon attractor   (b) The Holmes attractor

FIGURE 1.5:   The Hénon and Holmes attractors

Another model, with very interesting properties, was proposed by Holmes (Holmes, 1979a):

$$x_{t+1} = bx_{t-1} + ax_t - x_t^3$$
(1.10)

Fig 1.5(b) presents the Holmes model when the parameters take the values $a = 3, b = 0.8$.

The Hénon model, along with other related models, extensions and variations, such as the *Holmes* model, are presented in Chapter 5.

1.3   More Complex Structures

1.3.1   Three-dimensional and higher-dimensional models

A number of interesting three-dimensional models are studied in Chapter 6. These include the Lorenz (Lorenz, 1963) and Rössler models, the autocatalytic and Arneodo (Arneodo et al., 1980, 1981b) models, and their extensions and variations. 3-dimensional views of the Lorenz and autocatalytic attractors are illustrated in Figures 1.6(a) and 1.6(b) respectively.

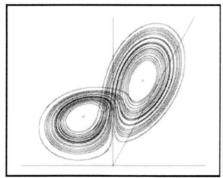
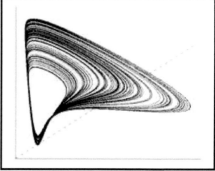

(a) The Lorenz attractor (3-d)   (b) The autocatalytic attractor

FIGURE 1.6:   The Lorenz and autocatalytic attractors

We also examine a four-parameter model of chemical reactions, which gives rise to a rich variety of chaotic patterns. A two-dimensional view of this chaotic attractor, along with the time development of one-dimensional oscillations, appears in Figure 1.7.

1.3.2   Conservative systems

In Chapter 7, we consider *conservative systems*, that is, systems that have a non-trivial first integral of motion, and in particular *Hamiltonian systems*. The existence of a first integral of motion provides enough information to draw the phase portrait of the system in cartesian coordinates. The resulting patterns exhibit a varying degree of symmetry, depending on the equations of the system. Figure 1.8 shows an example of two of the forms analysed in this chapter.

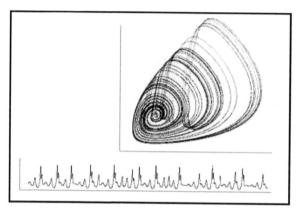

FIGURE 1.7:   A chemical reaction attractor

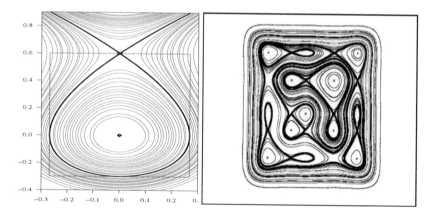

FIGURE 1.8:   Non-chaotic portraits

1.3.3   Rotations

Chapter 8 is devoted to rotations, reflections and translations, and the chaotic forms and attractors that these give rise to. The models in this category are characterized by the presence of rotations and translations, as well as non-linear terms, in the system equations. For appropriate values of the parameters, these models give rise to chaotic patterns and attractors. The most popular of these is probably the *Ikeda* model (Ikeda, 1979) (equations (1.14) and (1.13)).

Rotation models express a large variety of systems in various disciplines. A simple rotation in the plane, followed by a translation, does not in itself lead to chaotic forms. The rotation equation, in its usual form, is based on linear functions such as (1.11):

$$
\begin{aligned}
x_{t+1} &= x_t \cos(\theta_t) - y_t \sin(\theta_t) \\
y_{t+1} &= x_t \sin(\theta_t) + y_t \cos(\theta_t)
\end{aligned}
\tag{1.11}
$$

However, the introduction of non-linear terms in system (1.11) gives chaotic solutions. First of all, let us consider the effect of adding a quadratic term in both equations. The system then becomes non-linear:

$$
\begin{aligned}
x_{t+1} &= x_t \cos(\theta) - y_t \sin(\theta) + x_t^2 \sin(\theta) \\
y_{t+1} &= x_t \sin(\theta) + y_t \cos(\theta) - x_t^2 \cos(\theta)
\end{aligned}
\tag{1.12}
$$

When the rotation angle is stable, $\theta = 1.3$, the system gives an interesting two-dimensional map with chaotic regions and islands of stability inside these regions (Figure 1.9).

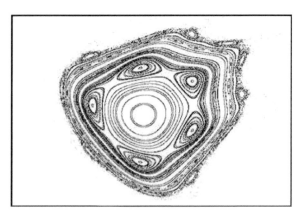

FIGURE 1.9:   A rotation chaotic map

Another way to add non-linearity is to assume that the rotation angle θ is a non-linear function, as is the case of the famous Ikeda attractor. The model is based on a system of two rotation equations with a translation parameter a added to the first equation, a space parameter b, and a non-linear rotation angle of the form:

$$\theta_t = c - \frac{d}{1 + x_t^2 + y_t^2} \tag{1.13}$$

The rotation-translation system in this case takes the following form:

$$
\begin{aligned}
x_{t+1} &= a + b(x_t \cos(\theta_t) - y_t \sin(\theta_t)) \\
y_{t+1} &= b(x_t \sin(\theta_t) + y_t \cos(\theta_t))
\end{aligned} \tag{1.14}
$$

This system was first proposed by Ikeda (1979) to explain the propagation of light into a ring cavity. The parameter values in that case where $a = 1, b = 0.83, c = 0.4$ and $d = 6$, and they give rise to a very interesting attractor illustrated in Figure 1.10.

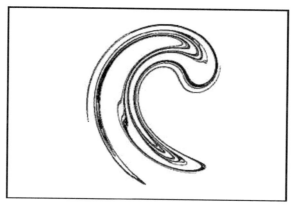

FIGURE 1.10: The Ikeda attractor

This map has Jacobian $J = b^2$. When $b < 1$ several chaotic and non-chaotic patterns can be seen in the various simulations, some of which are illustrated in Chapter 8.

The properties of the chaotic maps in this category are so rich and often unexpected. The shape of the Ikeda attractor in the previous standard form depends on the selection of the values of the parameters. By varying these parameters, very interesting new forms appear. A wide variety of these forms are presented and analysed in Chapter 8. Two of these attractors are illustrated in Figures 1.11(a) and 1.11(b). In Figure 1.11(a) an attractor pair appears. The parameters are $a = 6, b = 0.9, c = 3.1$ and $d = 6$. In Figure 1.11(b), only the parameter c is changed ($c = 2.22$). Now three images of the attractors are present.

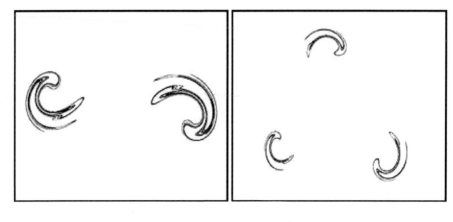

(a) A double Ikeda attractor   (b) A triple Ikeda attractor

FIGURE 1.11:   Ikeda attractors

1.3.4   Shape and form

In Chapter 9 we study how chaotic analysis can be used as a tool to design forms that appear in physical systems. Such an analysis is useful in the scientific field related to shape and form. Even the most strange chaotic attractors and forms can have real significance, especially in biological systems but also in fluid flow or in crystal formation. The analysis of these systems has moved from the classical analysis concerning simple and regular forms to chaotic analysis. The aim here is to go beyond the classical stability analysis. We try to find mechanisms to design simple or complicated and chaotic forms, and to pass from a qualitative to a quantitative point of view. The sea-shell formation illustrated earlier in this chapter is one case of an attractor with a special shape. The equations used are 3-dimensional and the original inspiration comes from the Rössler model.

The main tools for shape formation are *rotation* and *translation*, as in the Ikeda case above, but also *reflection with translation*, based on the following set of equations:

$$
\begin{aligned}
x_{t+1} &= a + b(x_t \cos(2\theta_t) + y_t \sin(2\theta_t)) \\
y_{t+1} &= \quad b(x_t \sin(2\theta_t) - y_t \cos(2\theta_t))
\end{aligned}
\tag{1.15}
$$

If the same function as in the Ikeda case is used for the angle θ_t

$$
\theta_t = c - \frac{d}{1 + x_t^2 + y_t^2}
$$

then we obtain an attractor (Figure 1.12(a)) which is divided in two new separate forms (Figure 1.12(b)). The parameters in the first case are $a = 2.5, b = 0.9, c = 1.6$ and $d = 9$. In the second case, $a = 6$ and the other parameters remain unchanged.

The centres of the two attractors are located at:

$$(x, y) = (a, 0) \text{ and}$$
$$(x, y) = (a + ab \cos(\theta), ab \sin(\theta))$$

where $\theta = c - \frac{d}{1+a^2}$. The reflection attractor is on the left side of Figure 1.12(b). These cases simulate the split of a vortex in two separated forms. The parameter a is quite higher than in the previous case.

(a) Reflection with 1 image   (b) Reflection with 2 images

FIGURE 1.12: Translation-reflection attractors

The rotation equation (1.14), applied to the Ikeda case, but now with $b = 1$, is a space-preserving transformation. Interesting forms arise assuming that the rotation angle is a function of the distance $r_t = \sqrt{x_t^2 + y_t^2}$. For instance, when

$$\theta_t = c + \frac{d}{r_t^2}$$

we obtain the form illustrated in Figure 1.13(a). There is a symmetry axis at $x = \frac{a}{2}$ and one equilibrium point in the central part of the figure, located at $(x, y_{t+1}) = \left(\frac{a}{2}, y_t\right)$. The five equilibrium points in the periphery are of the fifth order: $(x_{t+5}, y_{t+5}) = (x_t, y_t)$. One of these points is also located on the $x = \frac{a}{2}$ axis, whereas the other four are located in symmetric places about this axis. The parameters are $a = 3$, $c = 1$ and $d = 4$.

Furthermore, the Ikeda model with $b = 1$ provides interesting forms with the same axis of symmetry, $x = \frac{a}{2}$. Figure 1.13(b) illustrates this case with parameter values $a = 1$, $c = 5$ and $d = 5$. In the figure there are two pairs of second-order equilibrium points, five fifth-order and eight eight-order equilibrium points.

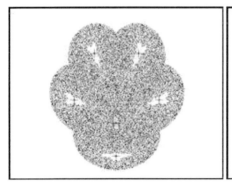
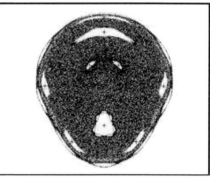

(a) Fifth-order symmetry   (b) Multiple-order symmetries

FIGURE 1.13:   Symmetric forms

In Chapter 9, several symmetric and non-symmetric forms are presented and analysed. The following figures illustrate a symmetric chaotic form before the separation step (Figures 1.14(a) and 1.14(b)), in the intermediate stage (Figure 1.14(c)), and after the separation of the chaotic form into two symmetric forms (Figure 1.14(d)). The parameters are $b = 1$, $c = 2.85$ and $d = 16$, and the parameter a takes the values $a = 18, 35, 45$ and 55 respectively. The rotation iterative procedure is used, with rotation angle $\theta_t = c + \frac{d}{r_t}$.

Another category of patterns includes forms which we call "flowers." By using the rotation formula with rotation angle

$$\theta_t = c + \frac{d}{r_t^\delta}$$

and with parameter values $b = 0.9$, $a = c = d = \frac{2\pi}{k}$, $k = \frac{17}{12}$ and $\delta = 2$, the graph in Figure 1.15(a) appears. When $a = \frac{2\pi}{3k}$, $c = d = \frac{2\pi}{k}$ and $\delta = 3$, Figure 1.15(b) results.

1.4   Chaos and the Universe

1.4.1   Chaos in the solar system

Complicated and even chaotic orbits and paths in the solar system are studied and simulated in Chapter 11. Objects in the solar system mainly follow circular or elliptic orbits around a centre of mass. In this case the simulation can follow a purely geometric approach. To simplify the problem the centre of mass is assumed to be located in the centre of the large mass. The attracting force is, of course, gravity, and

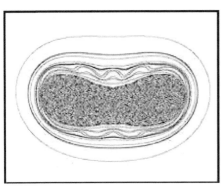

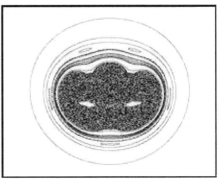

(a) One symmetric form   (b) Starting separation

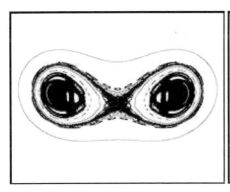

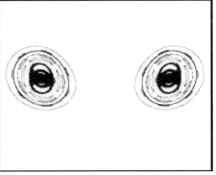

(c) First separation stage   (d) Final separation stage

FIGURE 1.14: Separation of chaotic forms

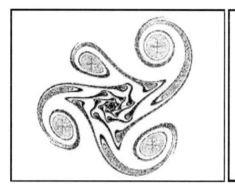 

(a) $\delta = 2$ (b) $\delta = 3$

FIGURE 1.15:   Flowers

is presented by an inverse square law of the distance. The non-dimensional equations
of motion for a body with small mass m rotating around a large body with mass M,
which we consider to stay in equilibrium position, are, in Cartesian coordinates:

$$\ddot{x} = -\frac{x}{(x^2 + y^2)^{3/2}}$$
$$\ddot{y} = -\frac{y}{(x^2 + y^2)^{3/2}} \tag{1.16}$$

The initial conditions are $x = 1$, $y = 0$, $\dot{x} = 0$ and $\dot{y} = v$. The escape velocity is
$v = \sqrt{2}$, and the circular orbit is obtained when $v = 1$. The circular orbit is illustrated
in Figure 1.16(a). In the same figure, an elliptic orbit is presented, where the initial
velocity is $1 < v = 1.2 < \sqrt{2}$. A hyperbolic orbit is also presented. The initial
velocity is higher than the escape limit $(v = 1.45 > \sqrt{2})$ and the particle flies off to
infinity. Figure 1.16(b) illustrates a large number of elliptic paths for the same value
of the initial velocity as before. The paths remain in the same plane, but deviate from
the original position, forming the shape presented in the figure.

The motions of two attracting point masses are relatively simple. The equations
of motion for the point mass m in Cartesian coordinates are:

$$\ddot{x}_1 = -\frac{(x_1 - x_2)m_2}{((x_1 - x_2)^2 + (y_1 - y_2)^2)^{3/2}}$$
$$\ddot{y}_1 = -\frac{(y_1 - y_2)m_2}{((x_1 - x_2)^2 + (y_1 - y_2)^2)^{3/2}} \tag{1.17}$$

Analogous is the case for the point mass m_2. The gravity constant is $G = 1$ and
the total mass of the rotating system is $M = m_1 + m_2$. If we give the two masses
initial positions (x_1, y_1) and (x_2, y_2), and initial velocities $\dot{x}_1$, $\dot{y}_1$, $\dot{x}_2$ and $\dot{y}_2$, then the

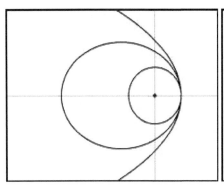
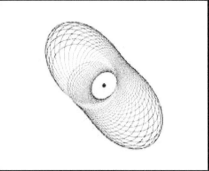

(a) Circular, Elliptic and Hyperbolic orbits   (b) Many elliptic orbits

FIGURE 1.16:   Orbits in two-body motion

masses revolve around each other while drifting together, as a pair, through space. Their centre of mass drifts at a constant velocity. Relative to an observer moving with the centre of mass, their orbits are ellipses and, in special cases, circles. The motion of two attracting points in space is presented in Figure 1.17(a). The heavy line represents the movement of the centre of mass. The masses revolve around the centre and drift together. The points have masses $m_1 = 0.4, m_2 = 0.6$ and initial velocities $\dot{x}_1 = -0.15$, $\dot{x}_2 = 0$, $\dot{y}_1 = -0.45$ and $\dot{y}_2 = 0.25$. The same motion, viewed relative to the centre of mass, is illustrated in Figure 1.17(b). Each point-mass follows an elliptic path.

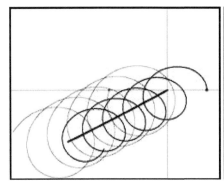
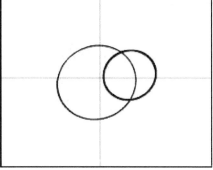

(a) Attracting bodies in space   (b) Relative movement

FIGURE 1.17:   Motion in the plane

The three-body problem, reduced to the plane, still gives very complicated and chaotic orbits. To simplify the problem we assume that the third body, a satellite, is of negligible mass relative to the other two bodies. The other two, the planets, travel around the centre of the combined mass in circular orbits in the same plane. The satellite rotates in the same plane. The total mass of the planets is $m_1 + m_2 = 1$. The equations of motion for the satellite then are:

$$\ddot{x} = -\frac{m_1(x - x_1)}{r_1^{3/2}} - \frac{m_2(x - x_2)}{r_2^{3/2}}$$

$$\ddot{y} = -\frac{m_1(y - y_1)}{r_1^{3/2}} - \frac{m_2(y - y_2)}{r_2^{3/2}}$$

(1.18)

where (x, y) is the position of the satellite, r_1, r_2 are the distances of the satellite from the two planets, m_1, m_2 are the masses of the two planets, and (x_1, y_1) and (x_2, y_2) are the positions of the planets at the same time t.

In the following example the planets have masses $m_1 = 0.9$, , $m_2 = 0.1$, and the position and initial velocity of the satellite are $(-1.033, 0)$ and $(0, 0.35)$ respectively. The complicated and chaotic paths are illustrated in the following figures. In Figure 1.18(a), the movement is viewed from the outside of the system, from space. It shows that the satellite can move between the two revolving planets. Figure 1.18(b) illustrates the orbit of the satellite viewed in a coordinate system that rotates with the revolution of the planets (rotating frame), with the X-axis always passing through the planets according to the rotation formulas:

$$X = x \cos t + y \sin t$$

$$Y = -x \sin t + y \cos t$$

(1.19)

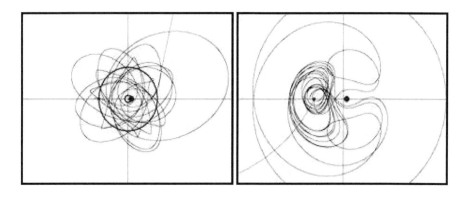

(a) Motion in space   (b) Motion in a rotating frame

FIGURE 1.18:   Motion in space

By using difference equations to model solar systems, it is possible to simulate chaotic patterns like the rings of Saturn and other planets. The simplest approach is to use an equation like the logistic, and to assume that the system rotates following elliptic orbits. Two realizations appear in the next figures. The parameter of the logistic model is $b = 3.6$ for Figure 1.19(a) and $b = 3.68$ for Figure 1.19(b). The rotation formula is:

$$x_t = 2r_t \cos t$$
$$y_t = 2r_t \sin t \tag{1.20}$$

where $r_{t+1} = br_t(1 - r_t)$.

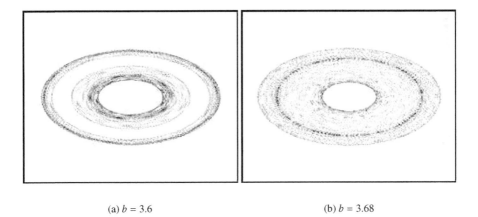

(a) $b = 3.6$ (b) $b = 3.68$

FIGURE 1.19: Ring systems

1.4.2   Chaos in galaxies

The underlying literature on galactic simulations deals mainly with the N-body problem and related simulations. The approach to the N-body problem follows some simplifications. The majority of the models we examined are based on numerical integration of the equations of motion for N mutually interacting particles with equal masses, which follow a steady rotational motion.[1] The acceleration of the i-th particle is:

$$\sum_{j \neq i}^{N} \frac{r_j - r_i}{\left(r_{ij}^2 + r_{cut}^2\right)^{3/2}}$$

---

[1] See Sellwood (1983, 1989); Sellwood and Wilkinson (1993).

The cutoff radius r_{cut} greatly simplifies the numerical computation by eliminating the infrequent, but very large, accelerations at close encounters. The gravity constant and the mass of each particle are set to 1. When N is large, the main computational problem is the large number ($\propto N(N-1)$) of operations required to determine the accelerations.

When a small perturbation is applied to the rotating system, familiar galaxy forms appear after some iterations. The details are presented in Chapter 11. Two characteristic examples of these forms are the two-armed spirals illustrated in Figures 1.20(a) and 1.20(b). The simulations start from an elliptic disk with different density distribution. In Figure 1.20(b), the high central density leads to the formulation of a barred spiral form.

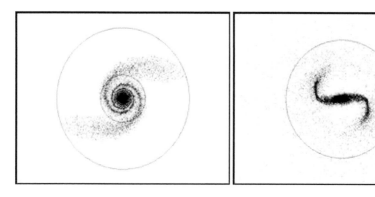

(a) Two-armed spiral $\left(\rho = \frac{c_2}{r^4}, \quad t = 100\right)$ (b) Barred spiral $\left(\rho = \frac{c_1}{r^8}, \quad t = 200\right)$

FIGURE 1.20:   Spiral galaxies

Another approach to producing a spiral pattern formation is based on the *elliptical epicycle* that a star follows around its guiding centre. According to the theory of kinematic spirals, the oval orbits are nested to form a two-armed spiral pattern of a galaxy.[2] The simulation results, illustrated in Figure 1.21(a), are based on the following equation in polar coordinates (r, θ):

$$r = \frac{R_g}{1 + a\cos(n(5 - 5R_g + \phi))}$$

The values of the guiding centre are $0.2 < R_g < 1$. The parameters are $a = 0.08$, $n = 2$ for the two-armed spiral and $a = 0.15$, $n = 1$ for the one-armed spiral (Figure 1.21(b)).

---

[2] See also Sparke and Gallagher (2000).

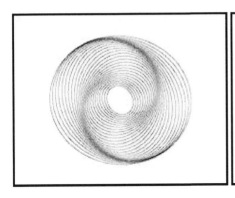 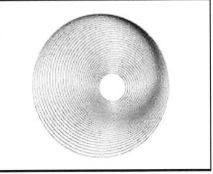

(a) Two-armed spiral   (b) One-armed spiral

FIGURE 1.21:   Spiral galaxies

Recent discoveries of new galaxies, intergalactic forms, planetary nebulae, black holes and other objects give inspiration for new simulations of the large variety of observed forms and shapes. The questions about black matter and how it influences the movement and shape of galaxies and other objects change our approach to model building and the related simulations of galactic forms. In many instances, the galactic forms arise by using very simple models and simulation tools. The forms presented below arise from the reflection iterative formula (1.20) with parameters $a = 1, b = 1$, and rotation angle:

$$\theta_t = c - \frac{d}{r_t^2}$$

The reflection procedure leads to mirror image symmetric shapes. In some instances a central chaotic bulge is created. Figure 1.22(a) illustrates the beginning of the mirror image formation, whereas in Figure 1.22(b) the central bulge is already present. The outer part has the form of an electromagnetic field. In both cases the parameter c is set to 1.6, while the parameter d is 35 for Figure 1.22(a) and 200 for Figure 1.22(b).

When $d = 800$, the reflection image takes the form presented in Figure 1.22(c). There is a central-bulge connected with the outer periphery by two symmetric routes. Figure 1.22(d) also illustrates a case where $d = 800$, but now $a = 0.6$. In this case, only the central bulge remains, while the form in the outer region is just beginning to take shape.

Rotations, reflections and translations with appropriate selection of the equation of the rotation angle give rise to a plethora of galaxy formations. Figure 1.23(a) illustrates a simple spiral pattern with one companion, whereas in Figure 1.23(b) two spiral galaxies with two companions appear. In both cases $b = 0.9, c = 2, d = 6$, and a is 0.3 for Figure 1.23(a) and 5 for Figure 1.23(b). The rotation angle in both

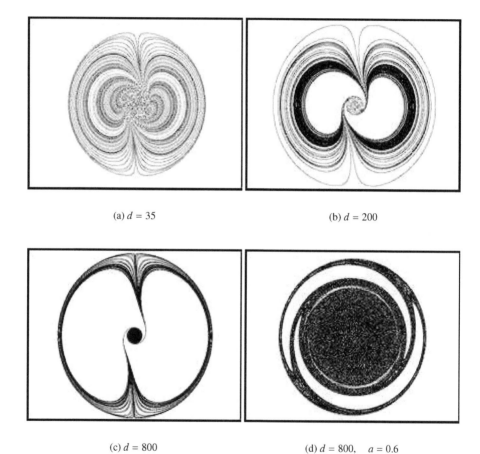

(a) $d = 35$

(b) $d = 200$

(c) $d = 800$

(d) $d = 800$, $a = 0.6$

FIGURE 1.22:   Reflection forms

case is given by:

$$\theta_t = c + \frac{d}{r_t}$$

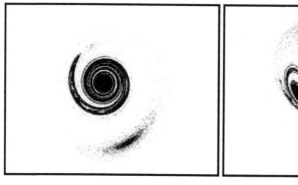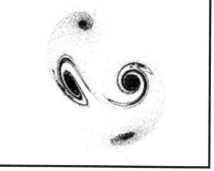

(a) $a = 0.3$ (b) $a = 5$

FIGURE 1.23: Spiral Patterns

1.4.3 Galactic-type potentials and the Hénon-Heiles system

Contopoulos (1958, 1960, 1965) explored the box-like paths of a body in a galactic potential, and found that the orbits in the meridian plane of an axisymmetric galaxy are like Lissajous figures. In his book (Contopoulos, 2002), Contopoulos explains that in 1958 he expected the orbits to be ergodic and fill all the space inside the energy surface. Instead, he found that the orbits did not fill all the available space, but instead filled curvilinear parallelograms, as in the case of deformed Lissajous figures. Later he could prove (Contopoulos, 1960) that such orbits can be explained both qualitatively and quantitatively by a formal third integral of motion.

The work of Hénon and Heiles (1964) on galactic motion is the first complete work that establishes the existence of chaos in galaxy formations. It provided considerable evidence of the existence of chaotic motions in galaxies. Their work started as a computer experiment in order to explore the existence and applicability of the third integral of galactic motion. Chapter 12 is devoted to this work of Hénon-Heiles and Contopoulos.

The works of Contopoulos and Hénon-Heiles are examples of what we will call *computer experiments*. This new type of experiments gave new directions in various scientific fields, and especially in astronomy. Sometimes the computer results were surprising and contradicted the existing theories of that time.

The Hamiltonian applied by Hénon and Heiles (1964) has the form

$$H = \frac{1}{2}(\dot{x}^2 + \dot{y}^2) + \frac{1}{2}\left(x^2 + y^2 + 2x^2y - \frac{2y^3}{3}\right) = h \tag{1.21}$$

So the potential $U(x,y)$ that is used has the form:

$$U(x,y) = \frac{1}{2}\left(x^2 + y^2 + 2x^2y - \frac{2y^3}{3}\right)$$

or in polar coordinates:

$$V(r,\theta) = \frac{1}{2}r^2 + \frac{1}{3}r^3 \sin(3\theta)$$

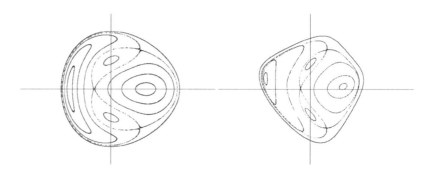

(a) The original Hénon-Heiles $(y, \dot{y})$ diagram

(b) A discrete alternative $(E = 1/12)$

FIGURE 1.24: Hénon-Heiles systems

We have reproduced here, via computer simulation, the original figure Figure 1.24(a) presented in the Hénon-Heiles paper. Hénon and Heiles estimated the paths of a particle according to the Hamiltonian (1.21) at the energy level $E = h = 1/12$, and then they provided the (y, y') diagram when $x = 0$. In this low energy level mainly regular orbits appear. At a higher energy level chaotic regions are formed.

A discrete alternative of the Hénon-Heiles model results in the image presented in Figure 1.24(b) for the energy level $E = 1/12$. In this case, $c = 0.1$. The discrete model is based on the following set of equations:

$$\begin{aligned}
x_{n+1} &= cx_n + u_n \\
y_{n+1} &= cy_n + v_n \\
u_{n+1} &= -x_n - 2x_{n+1}y_{n+1} \\
v_{n+1} &= -y_n - x_{n+1}^2 + y_{n+1}^2
\end{aligned} \tag{1.22}$$

The Jacobian of the system (1.22) is:

$$J = 1 + 2(c - 1)y_{n+1} + 4(c - 1)\left(x_{n+1}^2 + y_{n+1}^2\right)$$

It is obvious that for $c = 1$, J is 1, and therefore the space-preserving property is satisfied. However, when c takes values less than 1, very interesting figures arise ($c = 0.1$ in Figure 1.24(b)). The similarities with the Hénon-Heiles system (Figure 1.24(a)) are obvious.

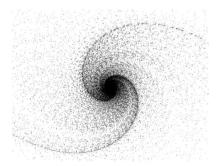

FIGURE 1.25: A two-armed spiral galaxy

Two-armed spiral galaxy forms can be obtained from simulations of very simple models. The two-armed spiral galaxy in Figure 1.25 is produced by using the following rotation model:

$$\begin{aligned} x_{n+1} &= b(x_n \cos a - y_n \sin a) + x_n^2 \sin a \\ y_{n+1} &= b(x_n \sin a + y_n \cos a) - x_n^2 \cos a \end{aligned} \tag{1.23}$$

In this case, $a = 0.2$ and $b = 0.9$. Since the Jacobian is $J = b^2$, b is an area contracting parameter.

The Hénon-Heiles system gives also interesting paths in the (x, y) plane. The paths are characterized by the level of potential. The form of the potential is triangular. Totally chaotic trajectories is illustrated in Figures 1.26(a) and 1.26(b). The energy level in both cases is $E = h = 1/6$, which is the escape limit. For this limit the equipotential triangle is drawn. All the (x, y) paths of the particle will be restricted inside the triangle. The initial values are $(x, y) = (0, 0)$, and $v = 0.1$. The value of u, 0.5668627, is estimated so that the value of the Hamiltonian is $h = 1/6$.

1.4.4 The Contopoulos system

A Hamiltonian with a "galactic-type" potential was first introduced by Contopoulos (1958, 1960) in his pioneering work on galaxies. The potential is based on the addition of two harmonic oscillators, along with higher order terms. The resulting

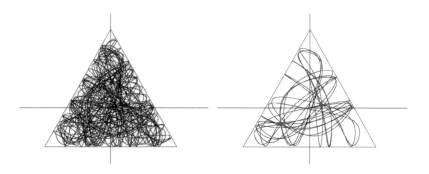

(a) A chaotic path over a long time period   (b) A chaotic path over a short time period

FIGURE 1.26:   Chaotic paths in the Hénon-Heiles system ($E = 1/6$)

form is

$$U(x, y) = \frac{1}{2} \left(w_1^2 x^2 + w_2^2 y^2 \right) - exy^2$$

with corresponding Hamiltonian:

$$H = \frac{1}{2} \left(u^2 + v^2 \right) + U(x, y) = h$$

Without loss of generality this Hamiltonian can be simplified to give the form:

$$H = \frac{1}{2} \left(u^2 + v^2 \right) + \frac{1}{2} \left(k^2 x^2 + y^2 \right) - exy^2 = \frac{1}{2}$$

where u, v, x, y, e have been rescaled, and $k = w_1/w_2$ is the important *resonance ratio*. The equations of motion are given by:

$$
\begin{aligned}
\dot{u} &= -\frac{\partial U}{\partial x} = -k^2 x - 2ey^2 \\
\dot{v} &= -\frac{\partial U}{\partial y} = \quad y - 2exy
\end{aligned}
\tag{1.24}
$$

   Resonant orbits, characterized as 4/1 (left image) and 2/3 (right image), are illustrated in Figure 1.27. For the 4/1 case the parameters are $e = 0.1$ and $k = w_1/w_2 = 4/1$, and the initial conditions are $x = 0.2$, $y = 0$, $u = 0$ and $v = 0.6$. For the 2/3 case the parameters are $e = 0.1$ and $k = w_1/w_2 = 2/3$, and the initial conditions are $x = 0.1$, $y = 0$, $u = 0$ and $v = 0.9977753$.

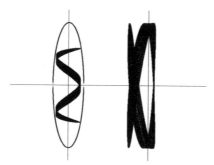

FIGURE 1.27:   The Contopoulos system: orbits and rotation forms at resonance ratio 4/1 (left) and 2/3 (right)

1.5   Odds and Ends, and Milestones

The book ends with Chapter 13, which is a collection of interesting systems of or variations on systems, including the effect of introducing noise into models, and an extensive discussion of the very interesting Lotka-Volterra system. The interesting images of Section 13.6 could certainly be considered as works of art. We hope they will act as an inspiration for future researchers.

The history of chaotic modelling is quite complex. A "milestones" table in Chapter 14 should provide a useful reference for future reading.

Chapter 2

Models and Modelling

2.1 Introduction

For many years scientists from all fields have been trying to describe the real world by constructing logical forms of varying complexity, called *models*. The process of selecting the main characteristics of a real life phenomenon and finding the basic mechanisms that describe the development of the phenomenon is called *modelling*, or also *model building*. Since the beginning of time, humans have been using models to understand the real world. The development of language enabled man to form important verbal constructs. These verbal constructs became the first models, expressing not just real objects, but also abstract ideas and innovations. Ancient Greeks established the first schools and academies, in order to encourage learning and to enhance the dissemination of information and exchange of ideas on scientific model building; a process known as dialectics, or more commonly *philosophy*.

Mathematical model building was also studied in Ancient Greece, mainly as a subset of geometry. Calculus and analysis came many centuries later. Nevertheless, their foundations can be found in the Pythagorean theory of fractions and in the Archimedean theory of polygonal approximations. Indeed Newton's "Method of Fluxions," where his treatment of calculus is presented, also includes a section on the "Applications on the Geometry of Curve-Lines." His explanation of Kepler's second law of planetary motion, as described in his famous letter to Halley,[1] is reminiscent of Archimedes's computation of the area of the circle by inscribed polygons. The paper, titled "De motu corporum in gyrum," is based on a purely geometric formulation of a model describing planetary motion.

In this book we analyse models that are expressed in mathematical terms. All the models we will examine fall under two categories, *differential equation models* and *difference equation models*, also called *maps*. During the modelling process, the standard methods of linear and non-linear analysis are extensively used. We pay particular attention to the methods of non-linear analysis based on *singular points*, *equilibrium points* and *characteristic trajectories*, and eigenvalues of the Jacobian, the characteristic matrix of the system.

---

[1] See the introduction in Acheson (1997).

2.2   Model Construction

Models are approximations of reality. In some cases, they describe the real situations quite well. However, the exact form of a model is not always possible to determine. The modelling process thus proceeds in two steps. First, a general form for the model is provided, typically in the form of a *modelling function*. This function will depend on some unknown parameters, initial conditions, and possibly other special characteristics, and thus can describe a number of different situations. The second step in the process is the estimation of these parameters so that the model approximates the real situation to the best of its potential a process typically known as "model fitting." In this section we will see numerous methods commonly used for constructing models.

2.2.1   Growth/decay models

An important general class of modelling functions are the so called *Growth Functions*. These are functions that express the growth of a system or of some special characteristics of a system, and they have, in general, a positive first time derivative. Similarly, functions with a negative first derivative would express the decline of a system, and would in general describe a decay process. The mathematical treatment in both cases is very similar.

It is important here to emphasise that one needs to strike a balance between the theoretical constructions of models, and the practical application of these models in real situations. Practical applications that lack a theoretical background do not contribute much to the development of any scientific field. A purely theoretical approach, on the other hand, becomes relevant only when supported by related applications (Skiadas, 1994; Skiadas et al., 1994).

Growth functions can be described in a number of different ways. We discuss some of these ways in the sections that follow.

2.2.1.1   Differential equation models

Models based on differential equations have been extensively used ever since the invention of calculus. In such models, the system under consideration is described by one or more functions that satisfy a system of differential equations. Thus, one doesn't always have a closed formula for the function, but a number of simulation and approximation techniques are available.

During the modelling process, one usually agrees upon the general form of the system, based on general considerations of the form of the solutions, and the expected behaviour of the system. One then proceeds to estimate the parameters involved, and to study the effect that these parameters have on the behaviour of the solution.

The most widely used growth/decay model is the exponential model, where the rate change of the system is proportional to the system's current state. This can be described by the simplest first order differential equation:

$$\dot{x} = bx \tag{2.1}$$

Here, as in the rest of this book, we will use Newton's notation, $\dot{x}$, for the *time derivative* of x:

$$\dot{x} \overset{\text{DFN}}{=} \frac{\partial x}{\partial t}$$

In equation (2.1), b is the model parameter that needs to be specified. Positive values of b will give rise to growth functions, while negative values will describe decay functions.

Sometimes systems such as (2.1) can be explicitly solved. In this particular case, as is typically shown in any calculus class, the solution is the familiar exponential function:

$$x(t) = x(0)e^{bt}$$

Often such an explicit description is not possible, and one has to rely on other quantitative and qualitative techniques for describing the solutions.

2.2.1.2 Difference equation models

Difference equations are the *discrete* analog of differential equations. The time parameter t is here assumed to be taking only discrete, values: $t = 0, 1, 2, \ldots$, corresponding to consecutive stages of the system, referred to as *steps*. The time derivative is then replaced by a *finite time difference*:

$$\dot{x} = \frac{\mathrm{d}x}{\mathrm{d}t} \approx \frac{\Delta x}{\Delta t} = \frac{x_{t+1} - x_t}{(t+1) - t} = x_{t+1} - x_t \tag{2.2}$$

The "rate of change" thus becomes simply the change in the system from the one stage to the next.

A difference equation is then nothing more than an algebraic relation between x_t and its successors, $x_{t+1}, x_{t+2}, \ldots$, or if you prefer of x_{t+1} and its predecessors. These difference equations are referred to as *maps* in the chaotic literature. For instance, equation (2.1) has the following discrete analog:

$$x_{t+1} - x_t = bx_t$$

or more simply:

$$x_{t+1} = bx_t \tag{2.3}$$

In difference equation systems, closed form solutions can sometimes be obtained by recursion and induction. For equation (2.3), this can be readily achieved and the solution is

$$x_t = x_0 b^t \tag{2.4}$$

which, as you will notice, greatly resembles the differential equation solution.[2]

2.2.1.3   Stochastic differential equations

The growth or decline of a system expressed by a variable x_t over time can often be formulated by two components:

1. The growth part or *infinitesimal growth*, $\mu(x, t)\, dt$

2. A measure expressing the *infinitesimal fluctuations*, $\sigma(x, t)\, dw_t$

A combination of these two components provides a *stochastic differential equation* of the general form:

$$dx(t) = \mu(x, t)\, dt + \sigma(x, t)\, dw_t \tag{2.5}$$

where the growth function $x = x(t)$ can, without loss of generality, be considered to be bounded $(0 \le x(t) \le 1)$,[3] the functions $\mu(x, t)$, $\sigma(x, t)$ are to be specified, and w_t is the standard Wiener process (Wiener, 1930, 1938, 1949, 1958).
    Every stochastic differential equation has a deterministic analogue, where the fluctuations are assumed to be zero:

$$dx(t) = \mu(x, t)dt$$

In many applications, the growth rate $\dot{x} = \frac{dx}{dt}$ is a function only of the magnitude x of the system, and not of time. Hence the deterministic analogue takes a much simpler form:

$$dx(t) = \mu(x)dt$$

2.3   Modelling Techniques

We will now discuss several general techniques for model construction. *Series approximation* (section 2.3.1) can help us reduce a complex system to a much simpler one that closely resembles it. It is a technique that we will use many times throughout the book.
    The most common source of models is the close investigation of real situations, and we discuss this process in section 2.3.2. In section 2.3.3, the *Calculus of Variations* approach is discussed, with its source in the Lagrangian reformulation of clas-

---

[2] In fact, if we keep in mind that the b in (2.3) here corresponds to the $1+b$ of (2.1), we can rewrite (2.4) as

$$x_t = x_0(1 + b)^t = x_0 e^{t \ln(1+b)} \approx x_0 e^{bt}$$

for small values of b.

[3] Since the system can be rescaled, so that $x(t)$ measures the percentage of the current state of the system over its "maximum state."

sical mechanics. Diffusion equations are discussed in section 2.3.4. Finally, section 2.3.5 discusses *delay differential equations*, that have the ability to express the delay in propagation of changes inherent in many natural and social systems.

2.3.1 Series approximation

When the function x_t describing a system takes values close to a particular point, then one can approximate a complex model by expanding the functions in the system in a Taylor series around that point. This method often gives sufficient results when the original model is simple, smooth and continuous, and the derivatives of the general equation of the model exist. As an example, consider a differential system of the form:

$$\dot{x} = f(x) \tag{2.6}$$

where $f(x)$ is a reasonably behaving function of x. If we are interested in values of x near 0, then we can replace f with the first couple of terms of its Taylor expansion:

$$f(x) = a_0 + a_1 x + a_2 x^2 + a_3 x^3 + a_4 x^4 + \cdots \tag{2.7}$$

Here the coefficients of the expansion can be readily interpreted as the derivatives of f at 0:

$$a_n = \frac{f^{(n)}(0)}{n!}$$

As an example of this approximation process, if we replace f in (2.6) by its linear approximation, and possibly after replacing x by $x + c$, we obtain the exponential model:

$$\dot{x} = a_1 x$$

In the majority of cases, using only the first (direction), and possibly second (curvature), derivative terms provides an adequate approximation to f, and hence to the model. Thus, it is not surprising that systems where f is a quadratic polynomial approximate very well other more complicated systems. Of course, if necessary, the higher terms of the Taylor expansion may also be used.

Often even this quadratic form can be simplified further. Symmetry of the system, quite common in physical systems, often leads to important simplifications of the approximation, and can lead directly to an expression of the model equation. For instance, if the x-space is isotropic,[4] we are led to the functional equation:

$$f(x) = f(-x)$$

Since the odd derivatives of an even function vanish at 0, the expansion (2.7), can be simplified to:

$$f(x) = a_0 + a_2 x^2 + a_4 x^4 + \cdots \tag{2.8}$$

---

[4]*i.e.* f is symmetric around the origin.

Probably the most well-known example of this type is the equation for the kinetic energy of a mass. If $f(v)$ represents the energy of the system, then a good approximation for small v (close to 0) would be:

$$f(v) = a_0 + a_2 v^2$$

The two terms can be identified physically as the *energy* at rest

$$f(0) = a_0$$

and the *kinetic energy*

$$f_{\text{kin}} = a_2 v^2 = \frac{1}{2} m v^2$$

On the other hand, economic, social and biological phenomena do not have this isotropic property. The same lack of symmetry holds for the time space, and the relevant Taylor approximations tend to include a linear term as well. The constant term in such a system can often be assumed to be zero.[5] Hence, a simple approximation will involve linear and quadratic terms. This approximation gives rise to a differential equation known as "The Logistic," first proposed by Verhulst (1845) to model the population growth in France. Later on, Pearl and Reed (1920) applied this model to express the population growth in the United States.

The logistic differential equation has the simple form

$$\dot{x} = bx(1 - x) \tag{2.9}$$

where x is the population at the present time, over the maximum level that the population could reach in the future. In other words, x is the *saturation level*.

To see how equation (2.9) follows naturally from (2.6) through the process of a Taylor series approximation, consider that x is restricted in the interval $[0, 1]$. We can consider the value $x = \frac{1}{2}$ where the population is at half its potential as the centre, and assume that the system will exhibit symmetry around $x = \frac{1}{2}$. In that case, the Taylor approximation to the second order would be:

$$f(x) = a_0 + a_2 \left(x - \frac{1}{2} \right)^2 = \left(a_0 + \frac{1}{4} \right) - a_2 \left(x - x^2 \right) = a + bx(1 - x)$$

Here a is the rate of growth of the population when the population is at $x = 0$ or $x = 1$, so we can reasonably assume that $a = 0$. The other parameter, b, is related to the growth rate at $x = \frac{1}{2}$. We are thus led to equation (2.9).

Equation (2.9) is a *separable differential equation*, and so can be easily solved explicitly, by rewriting it as

$$\frac{1}{x(1 - x)} dx = b dt$$

---

[5]For instance in birth processes or innovation diffusion cases.

The solution then would satisfy:

$$- \ln(1 - x) + \ln(x) = bt + C$$

This last equation can be rewritten in the final form:

$$x(t) = \frac{x_0 e^{bt}}{1 + x_0 (e^{bt} - 1)} \tag{2.10}$$

The graph of (2.10) exhibits a very characteristic smooth *sigmoid form* (Figure 2.1).

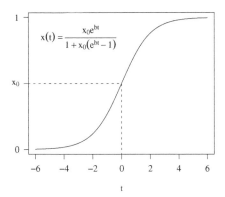

FIGURE 2.1: The solution to the logistic equation

A difference equation whose form closely parallels that of equation 2.9, but whose behaviour is a lot more interesting, also goes under the name "logistic":

$$x_{t+1} = bx_t (1 - x_t) \tag{2.11}$$

The behaviour of the solutions to (2.11) depends on the value of the parameter b. When $b < 3$, they follow a discrete sigmoid form. However, when $b > 3$ chaotic oscillations appear. The oscillations become progressively more chaotic as b approaches 4. The behaviour of the solutions of the discrete logistic will be explored further in Chapter 3.

The correct difference equation analogue to (2.11), using the approximation of the first derivative by a finite difference as in (2.2), is:

$$x_{t+1} = x_t + bx_t(1 - x_t)$$

Various second, third and higher order forms of differential and difference equations models can be constructed with similar methods as above. These equations may in some cases exhibit chaotic behaviour. These models will be examined more closely in subsequent chapters.

2.3.2 Empirical model formulation

In many cases in Physics, Biology, Economics, Marketing, Forecasting, *etc*, the construction of an appropriate general model follows by the empirical investigation of realistic situations. For instance, when the growth rate $\dot{x}$ is assumed to be proportional to the magnitude x of the system, then we end up with the exponential model discussed on page 31. Another example is based on the *relative growth rate*

$$\frac{\dot{x}}{x} = \frac{d}{dt}\ln x = (\ln x)' = \dot{y}$$

where y, the *relative magnitude* of the system, is defined as

$$y = \int \frac{dx}{x} = \ln x$$

and can be approximated by

$$y \approx \sum \frac{\Delta x}{x}$$

That is, the relative magnitude y is a measure of all the relative changes of the system from its original formation until its current status.

In this case, the standard measure of growth is the relative magnitude of the system, which is assumed as the metric scale of the system. This assumption leads to a convenient model construction especially when systems measured in completely different scales are to be compared. Examples of using the relative magnitude of the system as the standard measure are the Gompertz, Gaussian and Gamma models, whose descriptions follow.

2.3.2.1 The Gompertz model

The simplest assumption about the relative growth rate is that it is constant. This leads us to the exponential growth model $\frac{\dot{x}}{x} = b$ discussed already. However, as it is reasonable to expect that a system has limits to its growth, it would be more realistic to assume a declining relative growth rate. Assuming that this decline is proportional to the relative growth y of the system, we are led to a simple differential equation in y:

$$\dot{y} = -by \tag{2.12}$$

or

$$(\ln x)' = -b \ln x \tag{2.13}$$

which can finally be rewritten as:

$$\dot{x} = -bx \ln x \tag{2.14}$$

This model was first proposed in Gompertz (1825). Direct integration gives as solution the *Gompertz function*:

$$x = e^{\ln(x_0)e^{-bt}}$$

Using the Taylor approximation methods of section 2.3.1, we can examine the Gompertz system near $x = 1$: The logarithm $\ln(x)$ can be approximated by:

$$\ln(x) \approx x - 1$$

Using this approximation in (2.14) would result exactly in the logistic model. So for x near 1, the Gompertz model can be approximated by the logistic model.

Another interesting variant occurs if we replace x by $1 - x$ in the Gompertz differential equation, resulting in:

$$\dot{x} = -b(1 - x)\ln(1 - x)$$

This model arises by considering the relative decay of the system, instead of the relative growth. This model has skewness opposite to that of the Gompertz model, as now the maximum growth rate is achieved when

$$x = 1 - \frac{1}{e}$$

instead of when $x = \frac{1}{e}$. A comparison of the two models is given in Figure 2.2. Both models are referred to in the literature as "the Gompertz model," with different disciplines preferring one model over the other. The second variant is favoured in the actuarial sciences, as it is more intimately related to mortality.

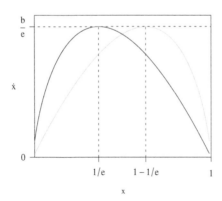

FIGURE 2.2: The two Gompertz models

2.3.2.2 The Gaussian and Gamma models

The well familiar Gaussian and Gamma distributions arise naturally from the relative-growth-rate setting, after some rather reasonable assumptions. For instance, if may assume that the relative growth rate is a linear function of time, which leads us to the following equation:

$$\dot{y} = (\ln x)' = a - bt \tag{2.15}$$

where $b > 0$.

The solution to (2.15) is a Gaussian function, namely:

$$x = ce^{-\frac{(t-\mu)^2}{2\sigma^2}}$$

where $\mu = \frac{a}{b}$, $\sigma^2 = b$, and the integration constant c is determined by normaliza-
tion.

If, on the other hand, one assumes a slightly different decreasing pattern for the
relative growth rate:

$$\dot{y} = (\ln x)' = a(\ln t)' - \lambda$$

then the solution is related to the *Gamma function*:

$$x = c(\lambda t)^a e^{-\lambda t}$$

2.3.3   The calculus of variations approach

Calculus of variations is concerned with optimising a functional:

$$I = \int L(\dot{x}, x, t)\,dt \tag{2.16}$$

where the functional L, called the *Lagrangian*, summarises the dynamics of the system in question, in terms of $\dot{x}$, x and t. The first step in the modelling process in this
case is the construction of an appropriate Lagrangian function. The solution to the
optimisation problem is then given by the Euler-Lagrange equation:

$$\frac{d}{dt}\left(\frac{\partial L}{\partial \dot{x}}\right) - \frac{\partial L}{\partial x} = 0 \tag{2.17}$$

As a specific example, suppose the Lagrangian consists of two basic components:

$$L(\dot{x}, x, t) = G(\dot{x}) + V(x)$$

where $G(\dot{x})$ expresses the forces of growth, and $V(x)$ expresses the limits of the
system. Then the Euler-Lagrange equation becomes:

$$\frac{d}{dt}\left(\frac{\partial G(\dot{x})}{\partial \dot{x}}\right) - \frac{\partial V(x)}{\partial x} = 0 \tag{2.18}$$

The form of the functions $G(\dot{x})$ and $V(x)$ is not always known. However, as the
integral (2.16) involves a summation of the infinitesimal parts of a path, it is in many
cases valid to approximate the functions $G(\dot{x})$ and $V(x)$ using Taylor series, especially
when the growth system is continuous and smooth.

Note further an interesting property of equation (2.18): The first order term of a Taylor series expansion of $G(\dot{x})$ does not contribute to the equation. Hence, the simplest expression for $G(\dot{x})$ is $\dot{x}^2$, and this is sufficient in many systems. This assumption leads to a Lagrangian of the form

$$L = \frac{1}{2}\dot{x}^2 + V(x)$$

Equation (2.17) yields, in this case, the following differential equation of growth:

$$\ddot{x} = \frac{\partial V}{\partial x} \tag{2.19}$$

The simplest interesting expression for V is $V(x) = \frac{1}{2}bx^2$. Equation (2.19) then takes a very familiar form:

$$\ddot{x} = bx$$

When $b > 0$, this model expresses a growth process, while when $b < 0$, it expresses an *oscillating process*.

If we consider the Lagrangian as a function of the relative growth rate of a system, $y = \ln x$, then the exact same analysis will yield the equation:

$$\ddot{y} = \frac{\partial V}{\partial y}$$

Assuming V is constant, $V(y) = c$, we are led to the exponential model:

$$\ddot{y} = 0 \Rightarrow \dot{y} = b \Rightarrow (\ln x)' = b \Rightarrow \dot{x} = bx$$

As we've seen before, when $b < 0$ this equation expresses a decay process.

2.3.4   The probabilistic-stochastic approach

A lot of models have their roots in probabilistic-stochastic considerations, which give rise to certain partial differential equations. These are usually *diffusion equations*. The simple diffusion equation (D is the *diffusion coefficient*)

$$\frac{\partial u}{\partial t} = D\frac{\partial^2 u}{\partial x^2}$$

and the Fokker-Planck equation

$$\frac{\partial u}{\partial t} = -\frac{\partial(au)}{\partial x} + D\frac{\partial^2 u}{\partial x^2}$$

are the basis for many models.

A *reaction-diffusion equation* has the general form

$$\frac{\partial u}{\partial t} = f(u) + D\nabla^2 u$$

where, as usual, the Laplacian ∇^2 is given by:

$$\nabla^2 \equiv \frac{\partial^2}{\partial x^2} + \frac{\partial^2}{\partial y^2} + \frac{\partial^2}{\partial z^2}$$

A classical model of this type is the so-called *FitzHugh-Nagumo model* (FitzHugh, 1961, 1969; Nagumo et al., 1962). The one-variable form of this model is given by

$$\dot{u} = u(1 - u^2) + K + D\nabla^2 u$$

where K is a parameter.

Simulation of models of this type gives rise to important chaotic cases of pattern formation. These models explain the development of chaotic waves in fluids and chemical kinetics.

2.3.5   Delay growth functions

Delay growth models assume that the rate of growth, $\dot{x} = \frac{dx}{dt}$, is a function not only of (x, t), but also of some earlier time $(t - T)$, where T is the *delay*. The general form for a continuous first-order delay growth model is:

$$\dot{x} = f(x, (t - T), t)$$

A very simple delay growth model is the following multiplicative process:

$$\dot{x} = bx_{t-T}g(x_t) \tag{2.20}$$

Equation (2.20) is not easy to handle. However, an approximation of the delay term is possible if we retain the first two terms of a Taylor series expansion at t:

$$x_{t-T} \approx x_t - T\dot{x} \tag{2.21}$$

Substituting (2.21) in (2.20) and rearranging terms, we obtain the system:

$$\dot{x} = \frac{bxg(x)}{1 + bTg(x)}$$

Depending on the the function $g(x)$, many different models arise. If, for instance, g is set to be the decreasing function $g(x) = 1 - x$, then we obtain the GRM1 model proposed in Skiadas (1985) to express an *innovation diffusion* process:

$$\dot{x} = \frac{bx(1 - x)}{\sigma + (1 - \sigma)x} \tag{2.22}$$

where $\sigma = 1 + bT$. When no delay is present, $\sigma = 1$ and (2.22) reduces to the logistic differential equation (2.9).

The discrete alternatives of delay growth models, expressed by difference equations, give chaotic solutions for certain parameter values. These are examined in some detail in Chapter 4.

2.4   Chaotic Analysis and Simulation

In the past decades, extensive studies in various fields led to the creation of a new field of analysis, called "Chaotic Analysis." The innovation of this new field lies in its multidisciplinary approach to the problems studied, and its use of various methods and tools not applied before. Most notable among these is the use of computers for the related analysis and simulation, that has become an integral part of Chaotic Analysis. Determining the number of equilibrium islands and their locations inside a chaotic sea is done first by using chaotic simulations, and only later by exact calculations, when possible. If we exclude computers and simulations from chaotic analysis, an essential part of the chaotic phenomena, "the essence of chaos" according to Lorenz (Lorenz, 1993), will be lost.

As an example of the advantages of simulation, consider the following formation of two variants of a two-dimensional sea urchin, by using the logistic difference equation and a simple rotation scheme. The equations used are:

$$\begin{aligned}
f_{t+1} &= bf_t(1 - f_t) \\
x_t &= (1 - f_t)\cos(\theta_t) \\
y_t &= (1 - f_t)\sin(\theta_t)
\end{aligned} \tag{2.23}$$

The rotation angle θ_t changes by $h = 0.01$ in each time step, and the chaotic parameter b is set to $b = 1 + \sqrt{8}$ in Figure 2.3(a) and $b = 1.1 + \sqrt{8}$ in Figure 2.3(b).

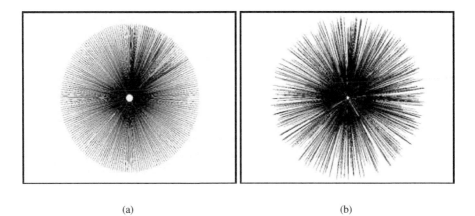

(a)   (b)

FIGURE 2.3:   Chaotic forms

The shape of a snail is formed from the same logistic difference equation as above, but now the chaotic band (t, f), which appears on the left side of Figure 2.4, rotates and simultaneously sinks, due to the addition of an exponential decay factor, in order to form the two-dimensional shape of a snail (right). The system of equations used is:

$$f_{t+1} = bf_t(1 - f_t)$$
$$x_t = (1 - f_t)\cos(\theta_t)e^{-at} \qquad (2.24)$$
$$y_t = (1 - f_t)\sin(\theta_t)e^{-at}$$

The chaotic parameter used here is $1.03 + \sqrt{8}$, and the time step is $h = 0.005$.

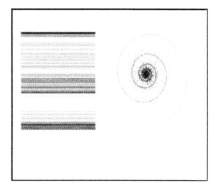

FIGURE 2.4:   A snail's pattern

2.5   Deterministic, Stochastic and Chaotic Models

Deterministic models are expressed by specific, deterministic equations, which give the same results for the same initial conditions. Stochastic models on the other hand are given by *stochastic equations*, which may give different paths, and different final results, for the same initial conditions. The chaotic models can act as deterministic models for some parameter values, but for some other values of the chaotic parameters they will give radically different final values for infinitesimally small changes in the initial values. Chaotic models are in a sense deterministic, in that the final values are determined completely from the precise initial values. From a practical point of view however, given the sensitivity of chaotic models to the initial conditions, the unpredictability of the final results and the non-uniqueness of the paths followed by the system in chaotic systems are in some ways similar to the unpredictability and non-uniqueness present in stochastic models.

As an illustration of this idea, one could attempt to use the logistic model as a random number generator. To achieve that, consider the one-dimensional stochastic process generated by the following set of difference equations:

$$x_{t+1} = 1 - 2x_t^2$$
$$y_{t+1} = y_t + ex_t \qquad (2.25)$$

Here, e is a small parameter ($e = 0.17$). Several paths of this process are illustrated in Figure 2.5(a). The resemblance of these paths to those of a Wiener process is striking. Another example, reminiscent of Brownian motion, appears in Figure 2.5(b). This figure shows a two-dimensional stochastic process based on a four-dimensional mapping.

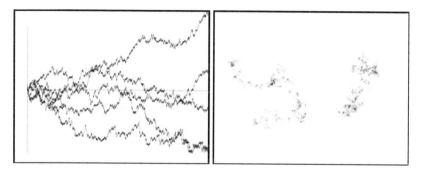

(a) One-dimensional diffusion   (b) Two-dimensional diffusion

FIGURE 2.5:   Chaos as randomness

Questions and Exercises

1. Construct the (t, x) graph for the map: $x_{t+1} = bx_t$. Compare this graph to the graph of the solution to the differential equation $\frac{dx}{dt} = bx$. Consider various values for b, including both negative and positive values, as well as values less than 1 and greater than 1.

2. Solve the differential equation $\frac{dx}{dt} = bx^2$, and graph the solution. Compare your results to the exponential (Malthus) model $\frac{dx}{dt} = bx$.

3. Solve the more general Exponential differential equation: $\frac{dx}{dt} = bx^a$, where the constant a is both positive and negative. Discuss the special cases resulting and draw graphs.

4. Provide examples and real-life applications modelled by the above two differential equations, including cases with positive as well as negative values for the parameters a, b.

5. For each of the following differential equations, construct both the "correct" and the "naive" difference equation analogue, and graph the (t, x) map. Compare this map with the solution to the differential equation.

    (a) $\frac{dx}{dt} = bx$

    (b) $\frac{dx}{dt} = bx^2$

    (c) $\frac{dx}{dt} = bx^2(1 - x)$

6. Using the Taylor expansion techniques, show that the differential equation model $\frac{dx}{dt} = \sin(bx)$ may be approximated by the exponential model when x is small. Use a simulation to compare the two models, and verify that the approximation is indeed good near $x = 0$.

7. Using the Taylor expansion techniques, what simple model do you find that approximates the model $\frac{dx}{dt} = \cos(bx)$ for small x?

8. Use the calculus of variations methods described in section 2.3.3 to explain why the very simple models described there, $\frac{dx}{dt} = bx$ and $\frac{d^2x}{dt^2} + bx = 0$, end up explaining very frequently real-life situations.

9. The second order differential equation $x'' + bx = 0$ provides oscillating solutions for positive values of the parameter b. Find the difference equation alternative of this differential equation and check if the oscillating character is retained in the map form. What is the result when $b = 1$?

10. What happens in the difference equation from the previous problem, but for negative values of b? What is the difference between the two cases?

11. Confirm that the maximum growth rates for the two Gompertz models are indeed at the points $\frac{1}{e}$ and $1 - \frac{1}{e}$ respectively. How much is the maximum growth rate?

Chapter 3

The Logistic Model

3.1   The Logistic Map

The *logistic map* is given by the *logistic difference equation*

$$x_{t+1} = bx_t(1 - x_t) \qquad (3.1)$$

Equation (3.1) has been extensively analysed in the last decades and has become the basis for numerous papers and studies. The analysis of this relatively simple iterative equation revealed a rich variety of inherent chaotic principles and rules. In particular, the study of the period doubling bifurcations that appear when the chaotic parameter b changes from one characteristic value to the next led to the formulation of a *universal law* in order to explain the phenomenon, at least for all quadratic maps (Feigenbaum, 1978, 1979, 1980a,d, 1983).

3.1.1   Geometric analysis of the logistic

We will present in this section the basic properties of the logistic map based on the classical analysis proposed by Feigenbaum and others. However, we will approach the subject from a more geometric perspective. In order to understand the evolution of (3.1), we consider the set of points

$$\{(x_t, y_t) | t = 1, 2, \ldots\}$$

where:

$$y_t = x_{t+1} = f(x_t) = bx_t(1 - x_t)$$

Thinking of the system (3.1) via this set of points allows us to examine the logistic system in a geometric way, by considering first of all the graphs of the functions $y = f(x) = bx(1 - x)$ and $y = x$. Given a value x_t, we locate $x_{t+1} = y_t = f(x_t)$ by locating the point on the graph of $y = f(x)$ with $x = x_t$. We then use the line $y = x$ to project this y coordinate back to the x axis (Figure 3.1(a)).

If we continue this process, we can trace the evolution of the system, as in Figure 3.1(b). A point of particular interest is the point where the graph of f meets the line $y = x$, with coordinates $x = y = 1 - \frac{1}{b}$. If x_t is the coordinate of that point, then $x_{t+1} = f(x_t) = x_t$, and therefore the system is stationary.

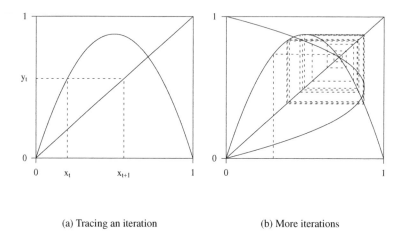

(a) Tracing an iteration   (b) More iterations

FIGURE 3.1:   Tracing iterations

Things get more interesting, and useful, when we consider other curves in place of $y = x$. Of particular importance is the inverse function to the logistic map, with equation:

$$x = f(y) = by(1 - y)$$

Suppose we follow steps similar to what we did above, except now we use this inverse logistic instead of the line $y = x$. Namely, starting from a value x_t, locate $y_t = f(x_t)$, and then use the inverse logistic to locate $x = f(y_t)$. The resulting point is $f(f(x_t))$; in other words it is x_{t+2}. This way, we can perform two iterations.

Suppose now that $(x, y) = (x_t, y_t)$ is a point on the intersection of the graphs of f and its inverse. Then we see that $x_{t+2} = x_t$ and $y_{t+2} = y_t$, so the system enters a two-point orbit. Algebraically, we would have the two equations:

$$y = f(x) = bx(1 - x)$$
$$x = f(y) = by(1 - y)$$

(3.2)

This system can be solved by observing that:

$$f(f(x)) = f(y) = x$$

or in our case:

$$x = b^2 x(1 - x)[1 - bx(1 - x)]$$

(3.3)

Equation (3.3) has four roots, corresponding to the four points where the curves of the logistic map and the inverse logistic map meet. Two of these will be the points where these maps meet the line $y = x$, namely the origin and the stationary point we discussed already. Let us consider for a moment the geometric significance of the other two points. If (x_t, y_t) is one of these points, then (x_{t+1}, y_{t+1}) will be the other one, and (x_{t+2}, y_{t+2}) will be the same as (x_t, y_t) and so on. In other words, the system will alternate between these two states. We will call these points *stationary of second order*. These are the first bifurcation points of the logistic map (Figure 3.2).

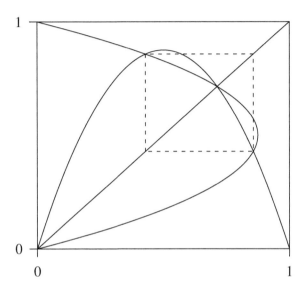

FIGURE 3.2:   Second order stationary points

Similarly, higher order bifurcation points can be found by considering the points of intersection of the iterates[1] of f, *i.e.* systems of the form:

$$y = f^k(x)$$
$$x = f^m(y)$$

(3.4)

Geometrically, tracing the first map followed by the second map results in transitioning from x_t to x_{t+k+m}. Algebraically, we would have to solve a polynomial of degree $2(k + m)$.

An illustrative example of the bifurcation of the logistic map is presented in Figure 3.3 when $b = 3.5$. The starting value is $x_0 = 0.1$. The final equilibrium is at one of the points where the logistic curve joins the inverse logistic curve.[2]

---

[1]The n-th iterate of f, denoted by f^k, is the function obtained by performing the function k times in succession. For instance, $f^2(x) = f(f(x))$, and in general $f^{n+1}(x) = f(f^n(x))$.

[2]In this example, the polygonal segments are related to f and its inverse, hence each subsequent x point is the result of two applications of f.

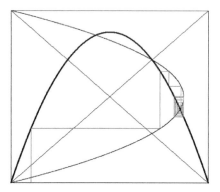

FIGURE 3.3: The logistic and the inverse logistic map for $b = 3.5$

Figure 3.4 illustrates a third order bifurcation cycle resulting when $b = 1 + \sqrt{8}$. The three points in the stationary orbit are the points of intersection of the quartic $y = f^2(x)$ with the inverse $x = f(y)$.[3]

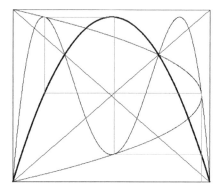

FIGURE 3.4: The third order cycle of the logistic and the inverse logistic map for $b = 1 + \sqrt{8}$

The value $b = 1 + \sqrt{8} \approx 3.828$ is the smallest value of b for which this third order orbit is present. Higher values of b quickly lead to chaotic behaviour. Figure 3.5 for instance presents a chaotic region for the logistic and the inverse logistic maps, when $b = 3.9$. A large number of repetitions is drawn.

---

[3]Those two curves have 5 points of intersection, but two of these points are the first order stationary points.

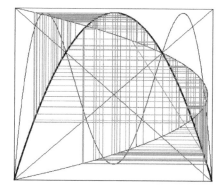

FIGURE 3.5: Chaotic region for the logistic and the inverse logistic map at $b = 3.9$

3.1.2 Algebraic analysis of the logistic

We will now proceed to analyse the behaviour of the logistic map from a more algebraic/analytic point of view. For a given value of b, the graph of the logistic is a parabola passing through the origin $(0, 0)$ and the point $(0, 1)$. The maximum of the parabola is obtained when $x = \frac{1}{2}$, and $y = f(x) = \frac{b}{4}$, as can be seen by calculating the first derivative of the logistic function:

$$f'(x) = b(1 - 2x)$$

and equating it to zero. Keeping in mind that $f(x)$ will be used as the x for the next iteration step, we see that we can only hope to constrain the system when $0 \leq b \leq 4$.

The map $y = f(x)$ has two fixed points, $(0, 0)$ and $\left(x_0 = 1 - \frac{1}{b}, 1 - \frac{1}{b}\right)$. The derivatives of f at the critical points are:

$$f'(0) = b \text{ and } f'(x_0) = 2 - b$$

We would like to know in which cases these points are *attracting*.[4] When x is close to 0, we have
$$f(x) = bx(1 - x) = bx - bx^2 \approx bx$$

so if x_t is close to 0, x_{t+1} is about equal to bx_t. Hence, if $b < 1$. x_{t+1} is even closer to 0, and consequently 0 will be an attracting point. The same is true when $b = 1$, since then $z_{t+1} = x_t - x_t^2 < x_t$. However, this is not the case any more when $b > 1$; x_{t+1} would tend to move away from 0.

Incidentally, $b > 1$ is exactly the range of values of b for which the second fixed point, $x_0 = 1 - \frac{1}{b}$, enters the picture, since it is now a point in the square $[0, 1] \times [0, 1]$.

---

[4]A set S of points is called attracting, if starting from any point sufficiently close to S the map stays close to S.

We will now carry a similar analysis at this point. A Taylor expansion of f at x_0 will provide us with some insight into the behaviour of the system near the point:[5]

$$f(x) = f(x_0) + f'(x_0)(x - x_0) + \frac{f''(x_0)}{2}(x - x_0)^2$$

Since we want to determine if the system will move towards x_0 or not, we consider the difference from x_0:

$$f(x) - x_0 = (f(x_0) - x_0) + f'(x_0)(x - x_0) + \frac{f''(x_0)}{2}(x - x_0)^2 \qquad (3.5)$$

We are interested in determining for which values of b, a small $x - x_0$ will result in an even smaller $f(x) - x_0$. When $x - x_0$ is small, the first term in (3.5), $f(x_0) - x_0$, will dominate the other two, if it is non-zero. so, in order for x_0 to be attracting, it is necessary that $f(x_0) - x_0 = 0$, i.e that x_0 is a fixed point. In that case, equation (3.5) becomes

$$f(x) - x_0 = f'(x_0)(x - x_0) + \frac{f''(x_0)}{2}(x - x_0)^2 \qquad (3.6)$$

For $x - x_0$ close to 0, the second order term will be negligible, and so:

$$f(x) - x_0 \equiv f'(x_0)(x - x_0)$$

Therefore, if $|f'(x_0)| < 1$, then x_0 is an attracting point, while if $|f'(x_0)| > 1$, then x_0 is not attracting. When $f'(x_0) = 1$, equation (3.5) becomes

$$f(x) - x_0 = (x - x_0) + \frac{f''(x_0)}{2}(x - x_0)^2$$

and the sign of $f''(x_0)$ determines whether x_0 is attracting or not.

In our case, the fixed point is $x_0 = 1 - \frac{1}{b}$, and the derivative, as computed earlier, is:

$$f'(x_0) = 2 - b$$

More generally, the Taylor expansion of f around x_0 in this case is:

$$f(x) - x_0 = (2 - b)(x - x_0) - b(x - x_0)^2$$

Consequently, x_0 is an attracting point for $1 \leq b < 3$. An example of the attraction process is illustrated in Figure 3.6 ($b = 2.9$, $x_0 = 0.07$).

When $b > 3$, the fixed point is no longer an attracting point, and a two-point, second order, attracting orbit appears. In order to obtain this orbit geometrically, the second order diagram (x_t, x_{t+2}), i.e. the graph of $f(f(x))$, is used, in addition to the (x_t, x_{t+1}) diagram, i.e. the graph of $y = f(x)$, and the (x_{t+1}, x_t) diagram, i.e. the graph of $x = f(y)$.

With a parameter of $b = 2.7$, we obtain Figure 3.7(a). It can be seen from the graph, that the only fixed points of $f(f(x))$ are the two fixed points of x. When

---

[5]Note that, since f is a polynomial of degree 2, f actually equals its second order Taylor expansion.

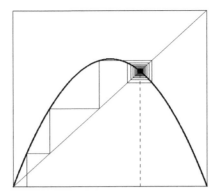

FIGURE 3.6: The logistic map

$b = 3$, all the curves pass through the point $\left(\frac{2}{3}, \frac{2}{3}\right)$ (Figure 3.7(b)). The slopes of the tangent lines to the curves at this point are -1 for $y = f(x)$ and $x = f(y)$, and 1 for $y = f(f(x))$.

When $b > 3$, we can compute the second order orbit by solving the equation

$$f(f(x)) = x$$

that is:

$$b^2x(1 - x)(1 - bx(1 - x)) = x \tag{3.7}$$

Equation (3.7) has four roots: The two fixed points of f ($x_1 = 0$ and $x_2 = 1 - \frac{1}{b}$), as well as two new roots, given by:

$$x_{3,4} = \frac{(b + 1) \pm \sqrt{(b + 1)(b - 3)}}{2b}$$

It is clear now why these points appear only when $b \geq 3$. This two-point orbit is attracting for $b > 3$, and remains attracting as long as b is such that the slope of the order-2 graph at the attracting points does not exceed -1. This leads to the equation:

$$b^2 - 2b - 5 = 0$$

which provides the upper value of b, $b = 1 + \sqrt{6} \approx 3.4495$, for which this two-point orbit is attracting. Figures 3.8(a) and 3.8(b) illustrate this order-2 bifurcation. Figure 3.8(a) shows the case where $b = 3.43$ and $x_0 = 0.01$. The order-2 bifurcation appears in the form of a square with corners located at the 4 characteristic points of the diagram. The tangent line to the order-2 graph at the larger of these points appears, with slope very close to -1. One can also clearly see that the fixed point of $y = f(x)$ is no longer an attracting point.

Another point to consider is the value of x at which the inverse of the logistic curve cuts the logistic curve at its maximum. This leads to the equation

$$b^3 - 4b^2 + 8 = 0$$

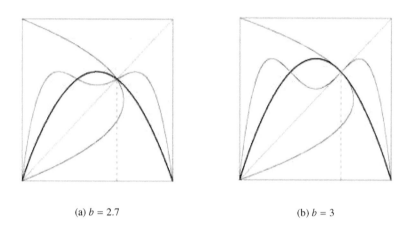

(a) $b = 2.7$ (b) $b = 3$

FIGURE 3.7:   The logistic map, $b \leq 3$

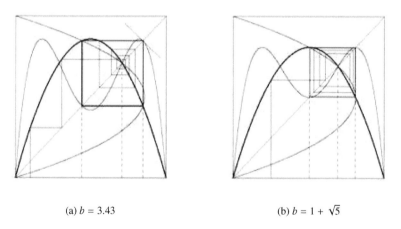

(a) $b = 3.43$ (b) $b = 1 + \sqrt{5}$

FIGURE 3.8:   The logistic map, $b > 3$

which has three real roots. Two of them are acceptable, $b = 2$ and $b = 1 + \sqrt{5}$. These are the same points where the order-2 curve meets the main diagonal.[6] Figure 3.8(b) illustrates the case where $b = 1 + \sqrt{5}$, with $x_0 = 0.25$.

The period-4 bifurcation appears in Figure 3.9. For better presentation, and in order to locate the characteristic points, we introduce the inverse (x_{t+3}, x_t) and (x_t, x_{t+4}) diagrams. The 4-point orbit can be thought of either as the points of intersection of the (x_t, x_{t+4}) diagram (purple line) with the line $y = x$, or as the points of intersection of the x_{t+3}, x_t diagram (blue line) with the curve $y = f(x)$. The parameter b is constrained in the range $1 + \sqrt{6} < b < 3.5440903 \cdots$. There are 4 characteristic values for x.

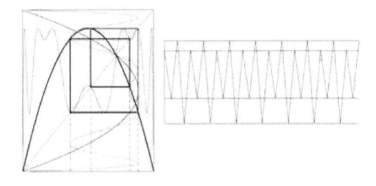

FIGURE 3.9: Order-4 bifurcation of the logistic map; real time and phase space diagrams

The right of Figure 3.9 is the real time diagram (x_t, t), whereas the left is the phase space diagram (x_t, x_{t+4}). In the real time diagram the order-4 oscillations appear. The values of x_t, after enough time has elapsed, oscillate between theses four limiting levels.

When $3.5440903 \cdots < b < 3.5644073 \cdots$ the period-8 bifurcation appears. Figure 3.10 illustrates this case. The chaotic parameter is $b = 3.5644$, and the inverse function is given by the (x_{t+7}, x_t) diagram. The order-8 logistic bifurcation is expressed by the (x_t, x_{t+8}) diagram. A real time diagram of the chaotic oscillations appears on the right of Figure 3.10.

The period-16 bifurcation is illustrated in Figure 3.11. The range covered by the parameter b is $3.5644073 \cdots < b < 3.5687594 \cdots$. For the application we set $b = 3.5687594 \cdots$. Figure 3.12 presents the period-32 bifurcation ($b = 3.56969$). The range for b is $3.5687594 \cdots < b < 3.5696916 \cdots$.

---

[6] Since if $y = f(x)$, then asking for $f(y) = x$ is the same as asking for $f(f(x)) = x$.

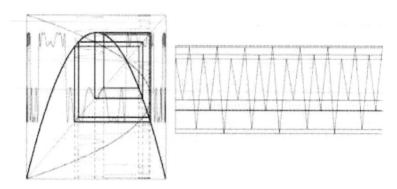

FIGURE 3.10:   Order-8 bifurcation of the logistic map; real time and phase
space diagrams

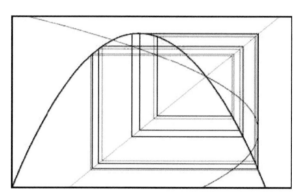

FIGURE 3.11:   Order-16 bifurcation and real time and phase space diagrams
of the logistic

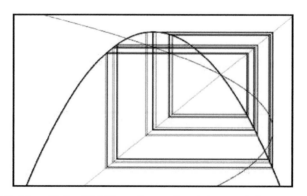

FIGURE 3.12:   Order-32 bifurcation and real time and phase space diagrams
of the logistic

The order-5 bifurcation appears in Figure 3.13. The inverse function is given by the (x_{t+4}, x_t) diagram. The order-5 logistic bifurcation is expressed by the (x_t, x_{t+5}) diagram. The chaotic oscillations appear in the real time diagram (right).

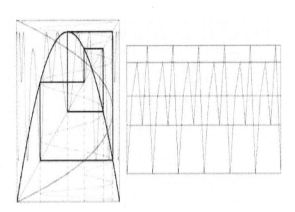

FIGURE 3.13: The order-5 bifurcation and real time and phase space diagrams

The order-6 bifurcation appears in Figure 3.14 ($b = 3.6267$). The inverse function is here given with the (x_{t+5}, x_t) diagram. The order-6 logistic bifurcation is expressed by the (x_t, x_{t+6}) diagram. The chaotic oscillations appear in the real time diagram (right).

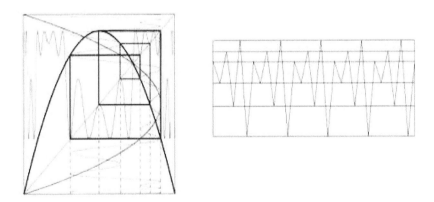

FIGURE 3.14: The order-6 bifurcation and real time and phase space diagrams

3.2   The Bifurcation Diagram

The period doubling and other properties of the logistic map are presented in the so-called *bifurcation diagram*. This is a two-dimensional graph (b, x_t) formed as follows: For each value of the chaotic parameter b, the values of x_t that form the stable orbit are plotted. Figure 3.15(a) shows how the bifurcation diagram looks for $b < 3.5$. For $0 < b < 1$, there is only one fixed point, at $x = 0$. For $1 < b < 3$, the fixed point is $x = 1 - \frac{1}{b}$. At $b = 3$, this splits up into the order-two orbit, which at 3.449 becomes unstable and splits up into an order-4 orbit and so on. The greyed out lines show the values that these orbits would have had if they were still stable.

The full bifurcation diagram for the logistic map appears in Figure 3.15(b). The first bifurcation point is located at $\left(b = 3, x = 1 - \frac{1}{b}\right)$, and the next two at $b \approx 3.4495$, the next four at $b \approx 3.54409$ and so on, according to the theory. These points can be calculated by solving the appropriate equation of the form $f^n(x) = x$, though this becomes exceedingly difficult as n increases. The construction of the bifurcation diagram typically follows these steps:

1. Break the range of b ($[0, 4]$) into small intervals.

2. For each interval, choose a b in that interval (say the left end-point), and perform repeated applications of the iterative process $x_{n+1} = f(x_n)$ until stability is achieved. At this point, the desired values x_n will be determined.

3. Move on to the next value of b. Use the previously found stable x values as starting points for the iteration process, to achieve stability much quicker.

The characteristic values of the parameter b at each bifurcation stage follow the *Feigenbaum law*. According to this law the sequence of period doubling, presented in the above figure by a bifurcation tree, is due to a quantitative convergence of the parameter b, which tends to a universal parameter $\delta = 4.6692016\cdots$. This parameter is given by the expression

$$\delta = \lim_{i \to \infty} \frac{b_{i-1} - b_i}{b_i - b_{i+1}}$$

The behaviour of the system can be understood better by considering a scaled version. More generally, if we consider a system

$$x_{n+1} = f(x_n)$$

then we consider the function:

$$g(x) = \lim_{n \to \infty} \frac{1}{f^{2^n}(0)} f^{2^n}\left(x f^{2^n}(0)\right)$$

Then, under some mild conditions on the map f, g satisfies the universal equation:

$$g(\alpha x) = -\alpha g(g(-x)) \tag{3.8}$$

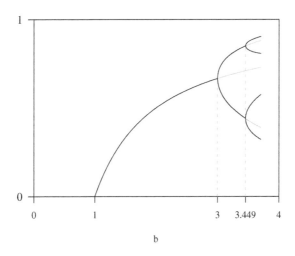

(a) The first two bifurcations

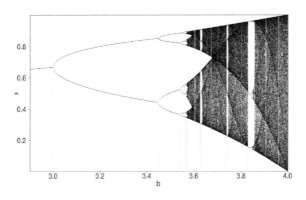

(b) The whole diagram

FIGURE 3.15:   The bifurcation diagram of the logistic

Here α is the second Feigenbaum constant, related to the ratio of the width between the prongs in a bifurcation, relative to the width in the previous bifurcation.

The function g can be approximated by a finite polynomial, and from that it is possible to estimate $\alpha = 2.50290787\cdots$, with the normalisation convention $g(0) = 1$ (Lanford, 1982a).

Other interesting properties arising from the bifurcation diagram come from observing the three "windows" which appear in a range of characteristic values of the parameter b around $b_6 = 3.627$, $b_5 = 3.74$ and $b_3 = 1 + \sqrt{8}$. These values of b give sixth, fifth and third order bifurcations respectively, and the related period doubling bifurcations lead to 12, 10 and 6 and higher order bifurcation forms. The period-3 bifurcation can be seen in Figure 3.16(a). In Figure 3.16(b) the order $2 \times 3 = 6$ bifurcation is illustrated ($b = 3.845$).

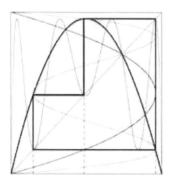   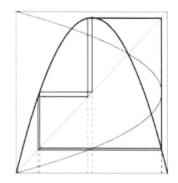

(a) Order-3 bifurcation   (b) Order 2×3 bifurcation

FIGURE 3.16:   Order-3 bifurcations

A change of variables

$$z = \frac{1}{2}b(1 - 2x)$$

results in the following expression for the logistic equation:

$$z_{t+1} = z_t^2 + c$$

where

$$c = \frac{b}{2} - \frac{b^2}{4}$$

From this point of view, the first period doubling occurs at $c = -\frac{3}{4}$, which corresponds to $b = 3$. The second period doubling occurs at $c = -\frac{5}{4}$, which corresponds to $b = 1 + \sqrt{6}$, while the period-3 solutions appear at a saddle-node bifurcation at

$c = -\frac{7}{4}$, corresponding to $b = 1 + \sqrt{8}$. An illustration of the period-3 bifurcation using the z-form of the logistic equation appears in Figure 3.17.

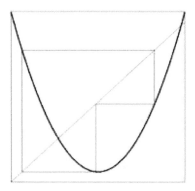

FIGURE 3.17: The order-3 bifurcation

3.3 Other Models with Similar Behaviour

Functions with varying equation forms show chaotic behaviour similar to the logistic. One of them is the *sinusoidal* map, with equation

$$x_{t+1} = b \sin(x_t)$$

where b is restricted to the interval $0 < b \leq \pi$. A more convenient form is:

$$x_{t+1} = b \sin(\pi x_t)$$

Now the interval for the parameter is $0 < b \leq 1$.

Another model is the Gompertz type model (Gompertz, 1825):

$$x_{t+1} = -bx_t \ln(x_t)$$

where $0 < b \leq e = 2.71829\cdots$.

Another model was proposed by May (May, 1972, 1974, 1976) to describe the population growth of a single species population, which is regulated by an epidemic disease at high density:

$$x_{t+1} = x_t e^{r(1-x_t)}$$

The chaotic behaviour of this model is similar to that of the logistic model, as illustrated by the bifurcation diagram (Figure 3.18, right). The order-4 bifurcation is presented on the left of Figure 3.18 ($r = 2.6$).

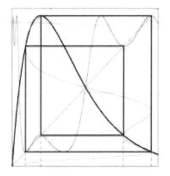 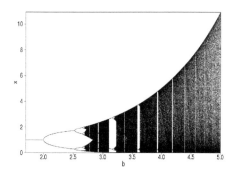

FIGURE 3.18:   The May model: an order-4 bifurcation (left) and the
bifurcation diagram (right)

3.4   Models with Different Chaotic Behaviour

Some Gaussian models show special chaotic behaviour, mainly due to the quadratic
exponent in the equations of these models. A simple Gaussian equation is:

$$x_{t+1} = b + e^{-ax_t^2}$$

The bifurcation diagram for $a = 4$ and $-1 < b < 1$ is shown on the left of
Figure 3.19(a). The model has two period doubling bifurcations, and then it returns
to the original state in a reverse order. A further increase in the parameter a of the
quadratic term, $a = 7.5$, gives a full sequence of period doubling bifurcations and
order-3 saddle-node bifurcations (Figure 3.19(b), right).

When the parameter a takes higher values, the chaotic behaviour of the model
changes radically. This is presented in Figure 3.20 ($a = 17$, $-1.2 < b < 1$).

A rich bifurcation diagram appears when the Gaussian model is combined with
the logistic to form the G-L model:

$$x_{t+1} = b - x_t(1 - x_t) + e^{-ax_t^2}$$

When $a = 7.5$, the bifurcation diagram in the range $-1.1 < b < 0.7$ has the
form illustrated in Figure 3.21. There are two main bifurcation regions, with period
doubling bifurcations and order-3 and order-5 saddle-node bifurcations. The order-5
bifurcation is presented in Figure 3.22(a) ($a = 7.5$, $b = 0.19$).

Figure 3.22(b) illustrates a perfect order-12 circle ($a = 7.5$, $b = 0.11$).

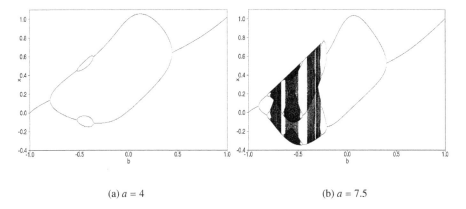

(a) $a = 4$ (b) $a = 7.5$

FIGURE 3.19:   Bifurcation diagrams of the Gaussian model

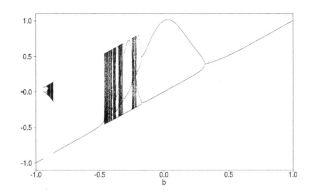

FIGURE 3.20:   Bifurcation diagram of the Gaussian model ($a = 17$)

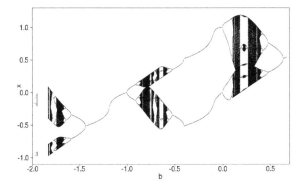

FIGURE 3.21:   Bifurcation diagram of the G-L model

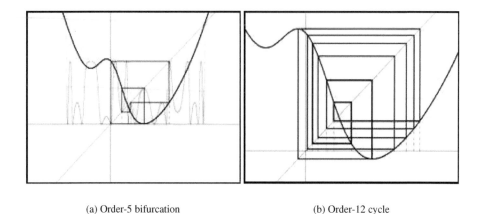

(a) Order-5 bifurcation   (b) Order-12 cycle

FIGURE 3.22:   Cycles in the G-L model

3.5   The GRM1 Chaotic Model

In this section we study the chaotic behaviour of the generalized rational (GRM1) innovation diffusion model. The deterministic continuous version of this model was proposed, analysed and applied in Skiadas (1985). Here, the chaotic behaviour is expressed through the discrete alternative to the continuous GRM1 model. The model shows symmetric and non-symmetric behaviour expressed by a parameter σ. When the *diffusion parameter* b and the σ parameter are in the range $b/\sigma \geq 2$, then the chaotic aspects of the model appear. A method is proposed for fitting the model to the data. Time series data expressing the cumulative percentage of steel produced by the oxygen process in various countries are used. Characteristic graphs of the chaotic behaviour are given and applications are presented.

3.5.1   GRM1 and innovation diffusion modelling

It has become commonplace to call this the information age, but an even more appropriate name might be the *innovation age*. The number of patent applications that the U.S. Patent and Trademark Office receives every year has been increasing exponentially since 1985, with the total number of patent applications exceeding $450,000$ in 2006 (Figure 3.23).

While not all patents translate to new products or new production methods, this growth clearly demonstrates a tendency, and this explosion of innovation activity presents significant challenges. Companies must have an appropriate way to describe the competitive dynamics in a market (Modis, 1997). Several *innovation diffusion models* have been presented, analysed and applied to real life data (Bass, 1969; Mahajan and Schoeman, 1977; Sharif and Kabir, 1976; Skiadas, 1985, 1986, 1987;

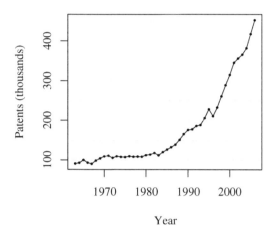

FIGURE 3.23: Patent applications in the United States

Modis and Debecker, 1992). A main direction of these applications focused on the non-symmetric behaviour of the models, expressed by specific parameters.

The GRM1 model is based on a family of *generalised rational models*. It was proposed as a relatively simple but very flexible model to express asymmetry during the innovation diffusion process (Skiadas, 1985, 1986). This model is expressed by the following differential equation

$$\dot{f} = b \frac{f(F - f)}{F - (1 - \sigma)f} \tag{3.9}$$

where f is the number of adopters at time t, F is the total number of potential adopters, b is the diffusion parameter, and σ is a dimensionless parameter. The GRM1 model has a point of inflection, varying from 0 to F as σ decreases from ∞ to 0. Another interesting property of σ is that it provides a measure of the asymmetry of the model. Perfect symmetry appears for $\sigma = 1$ when, equation (3.9) reduces to the logistic:

$$\dot{f} = bf\left(1 - \frac{f}{F}\right) \tag{3.10}$$

In the logistic model, bifurcation and further chaotic behaviour appear when $2 < b \leq 3$. However, various applications of the logistic model in several disciplines showed that the parameter b of the logistic model lies in very low limits, lower than 1. Thus, by using the logistic model it is not possible to express chaotic behaviour in real situations, as the estimated values of b fail to reach the limit at which chaotic behaviour appears. On the other hand, data provided for various cases shows that oscillations and chaotic behaviour appear quite frequently, and especially when the diffusion process is close to the upper limit F. Moreover, when the logistic model is applied in the form

$$x_{t+1} = bx_t(1 - x_t) \tag{3.11}$$

where $x_t = f_t/F$, then bifurcation and chaotic behaviour appear at even higher values of b, when $3 < b \le 4$.

As we show here, the GRM1 model exhibits chaotic behaviour for values of b that are quite low and are in accordance with the values estimated in real situations. This is accomplished with the help of the flexible parameter σ, which gives a measure of the asymmetry of the model. The chaotic behaviour of the model is analysed and illustrated by using appropriate graphs, especially (t, f) diagrams.

3.5.2 The generalized rational model

The discrete version of the continuous model (3.9) is given by:

$$f_{t+1} = f_t + b\frac{f_t(F - f_t)}{F - (1 - \sigma)f_t} \tag{3.12}$$

Some interesting properties of this model are illustrated in Figures 3.24(a) to 3.24(e).

In Figure 3.24(a), the proposed model takes the classical sigmoid form, whereas in Figure 3.24(b), a bifurcation appears as a simple oscillation. In Figure 3.24(c), a more complicated oscillation with four distinct oscillating levels appears, whereas in Figures 3.24(d), 3.24(e), and 3.24(f) a totally chaotic form appears. In all cases presented here the starting value is $f_0 = 1$, the upper limit $F = 100$, $b = 0.3$ and σ takes various values. The value selected for b is within the range 0.1 to 0.5, which is valid in real situations. By varying the dimensionless parameter σ, several forms of the model appear.

It is very important to consider the estimation of the values of the parameters b and σ for which bifurcation appears. The presence of the first oscillations and the onset of chaos, which follows, is a significant factor when studying innovation diffusion systems. According to the theory of chaotic models, bifurcation for the GRM1 model starts when:

$$f'_{t+1} = -1$$
$$f_{t+1} = f_t \tag{3.13}$$

Using equation (3.12), the resulting condition on the parameters b and σ is:

$$\frac{b}{\sigma} = 2$$

When $\frac{b}{\sigma} > 2$, oscillation and chaotic behaviour appear by gradually augmenting the fraction $\frac{b}{\sigma}$. When $\sigma = 1$, which is the case for the logistic model, bifurcation appears for $b > 2$.

It is also possible to obtain an analytical form for the values of f_t between the first two bifurcation points. To achieve this we consider the equation

$$f_{t+2} = f_t$$

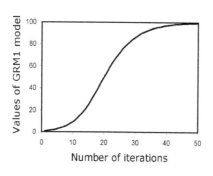

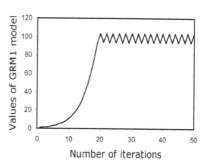

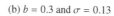

(a) $b = 0.3$ and $\sigma = 2$

(b) $b = 0.3$ and $\sigma = 0.13$

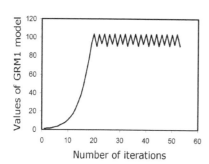

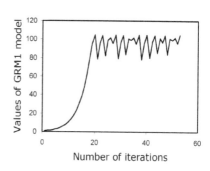

(c) $b = 0.3$ and $\sigma = 0.12$

(d) $b = 0.3$ and $\sigma = 0.10$

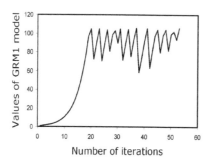

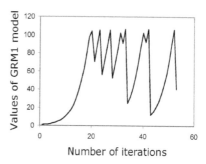

(e) $b = 0.3$ and $\sigma = 0.09$

(f) $b = 0.3$ and $\sigma = 0.08$

FIGURE 3.24: The GRM1 model

The exact formula of the solutions is:

$$f_t = F \frac{(b+2) \pm \sqrt{\frac{b(b+2)(b-2\sigma)}{(b-2\sigma+2)}}}{2(b+\sigma-1)} \tag{3.14}$$

For the logistic model ($\sigma = 1$) this reduces to:

$$f_t = F \frac{(b+2) \pm \sqrt{(b+2)(b-2)}}{2b} \tag{3.15}$$

When $b > 2\sigma$ in (3.14), or $b > 2$ in (3.15), the system oscillates at the values of f_t given by the above formulas respectively. When b is higher than these values, four distinct oscillating levels appear, and later eight and finally 2^n points. For sufficient specifically high values of b, n is very high and the system exhibits chaotic oscillations.

3.5.3   Parameter estimation for the GRM1 model

The parameters of the discrete GRM1 model are estimated by an iterative non-linear regression analysis algorithm by minimizing the sum of squared errors ($S = SSE$):

$$S = \sum \epsilon_t^2 = \sum_{t=1}^{n} (y_t - f_t)^2$$

where ϵ_t is the error term of the stochastic equation:

$$y_t = f_t + \sum_{i=1}^{n} \frac{\partial f_t}{\partial a_i} \Delta a_i + \epsilon_t$$

where y_t denotes the provided data and f_t is calculated for every t from the GRM1 equation, given a set of initial values for the parameters a_i. The estimation of parameters is highly sensitive to the presence of oscillations and chaotic oscillations in the provided data. For a better fit, it was decided to use the non-linear estimation method proposed by Nash for the discrete logistic model for only three parameters of the model, and fixing the dimensionless parameter σ. This parameter is gradually changed during the iterative procedure, until the sum of squared errors is minimised. The starting values for the partial derivatives need an estimation of the following forms given a set of initial values for the model parameters:

$$\frac{\partial f_1}{\partial b} = \frac{f_0(F - f_0)}{F - (1 - \sigma)f_0}$$

$$\frac{\partial f_1}{\partial f_0} = 1 + b \frac{F^2 - 2Ff_0 + (1 - \sigma)f_0^2}{(F - (1 - \sigma)f_0)^2} \tag{3.16}$$

$$\frac{\partial f_1}{\partial F} = b\sigma \left(\frac{f_0}{F}\right)^2$$

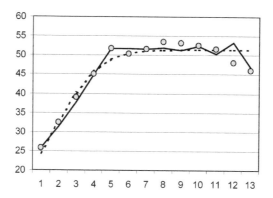

FIGURE 3.25: Spain, oxygen steel process (1968–1980)

Following the estimation of the initial values of the partial derivatives, the iterative procedure continues the estimation by using the formulae:

$$\frac{\partial f_{t+1}}{\partial b} = \frac{\partial f_t}{\partial b}(1 + bk_t) + \frac{f_t(F - f_t)}{F - (1 - \sigma)f_t}$$

$$\frac{\partial f_{t+1}}{\partial f_0} = \frac{\partial f_t}{\partial f_0}(1 + bk_t)$$ (3.17)

$$\frac{\partial f_{t+1}}{\partial F} = \frac{\partial f_t}{\partial F}(1 + bk_t) + \frac{b\sigma f_t^2}{(F - (1 - \sigma)f_t)^2}$$

where

$$k_t = \left(F^2 - 2Ff_t + \frac{(1 - \sigma)f_t}{F - (1 - \sigma)f_t}\right)^2$$

3.5.4 Illustrations

Time series data expressing the cumulative percentage of steel produced by the oxygen process[7] in various countries are used from Poznanski (1983). Figure 3.25 illustrates the diffusion of oxygen steel technology in Spain from 1968 to 1980, for a period of 13 years. The actual data includes 18 years, but it is more appropriate to study the last part of the time series, as this part shows the characteristic oscillations that are of special interest to us. The small circles indicate the actual data, the dotted line is the path of the logistic model and the solid line is the path of the GRM1 model.

Parameter estimates and the sum of squared errors are summarised in Table 3.1. The parameter b for the logistic model fit is relatively high, but is still far from the

---

[7]In this process, pure oxygen is introduced in order to burn the carbon inside melted steel, so as to improve the quality of the produced steel. Before the introduction of this method, air, consisting of a mixture of oxygen and nitrogen, was used instead.

TABLE 3.1:   Parameter estimates and sum of
squared errors (SSE) for logistic and GRM1 models in
Spain from 1968 to 1980

Model	b	l	F	$\sigma(b/\sigma)$	SSE
Logistic	0.6309	24.373	51.474	—	72.838
GRM1	0.2331	25.779	51.736	0.084 (2.775)	41.748

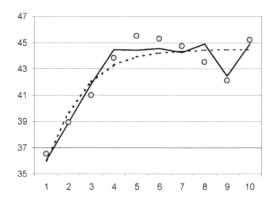

FIGURE 3.26:   Italy oxygen steel process (1970–1980)

value needed for the start of bifurcation ($b = 2$). The form of the logistic path
presented in the Figure 3.25 has a smooth form. The model fails to express the oscillating behaviour of the actual case studied. On the other hand, the GRM1 model
has a value for b lower than that of the logistic model, but the extra parameter
σ accounts for the presence of oscillating and chaotic behaviour, as the fraction
$b/\sigma = 2.775 > 2$. The estimated values for l and F are very close for both models. The ability of the GRM1 model to follow the oscillating behaviour of actual data
is illustrated in the Figure 3.25, and is also expressed by the strong improvement of
the sum of squared errors (SSE).

Figure 3.26 illustrates the diffusion of oxygen steel technology in Italy from 1970
to 1980. The process ends in an oscillating form. The discrete logistic fails to express
these oscillations, whereas the discrete GRM1 shows a considerable flexibility in
approximating the data. The sum of the squared errors is very low for the GRM1
model, compared to that of the logistic, as is shown in Table 3.2. The fraction $b/\sigma = 3.5292$ for the GRM1 model accounts for the chaotic behaviour.

Data for the diffusion of the oxygen steel process in Luxemburg are of considerable interest, as they cover a wide scale, from 1.5% during 1962 to 100% in 1980
(Figure 3.27). The GRM1 model showed a good flexibility, as it covers both the fast
growth at the first stages of the diffusion process, as well as the sudden turn to the
high platform of 100%. Also, the small fluctuations at the end of the process are
simulated quite well, as the fraction $b/\sigma = 3.609$ accounts for the chaotic region of

TABLE 3.2: Parameter estimates and sum of squared errors (SSE) for logistic and GRM1 models in Italy from 1970 to 1980

Model	b	l	F	$\sigma(b/\sigma)$	SSE
Logistic	0.5447	35.957	44.473	—	15.330
GRM1	0.08823	36.0402	44.4614	0.025 (3.5292)	7.431

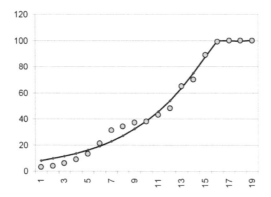

FIGURE 3.27: Luxemburg oxygen steel process (1962–1980)

the model. Figure 3.27 illustrates the case of Luxemburg for the following estimated values for the parameters: $b = 0.1931$, $l = 7.968$, $F = 99.669$ and $\sigma = 0.0535$. The mean squared error is $MSE = 20.872$.

The flexibility and ability of the GRM1 model to simulate growth processes that show oscillations, as well as chaotic oscillations, at the end of the process, are demonstrated in the following case of the diffusion of oxygen steel technology in Bulgaria from 1968 to 1978 (Figure 3.28). The estimates for the parameters are $b = 0.04046$, $l = 49.2425$, $F = 58.412$ and $\sigma = 0.012$. The sum of squared errors is $SSE = 21.431$ and the fraction $b/\sigma = 3.3718$ indicates that the model is in the region of chaotic behaviour.

3.6 Further Discussion

Several problems arise when applying the logistic map to real world data. Firstly, there are many cases where non-chaotic or even non-oscillating behaviour is present, even when the sigmoid logistic shape approaches the highest values, namely 1 = 100%. But, the analysis of the logistic map shows that bifurcation, and thus oscillat-

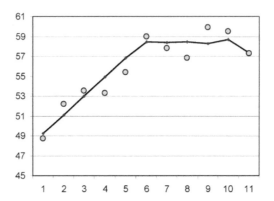

FIGURE 3.28:   Bulgaria oxygen steel process (1968–1978)

ing behaviour, starts at $b = 3$, which corresponds to $x = 1 - 1/b$, or $x = 2/3$ (66.66% level). If we follow the logistic model, all the processes exceeding this level must show oscillating or chaotic behaviour when approaching a higher level. However, this is not the case. Many logistic-like time-series data approach high levels without showing oscillating or chaotic behaviour.

The second problem when applying the logistic map to time-series data is that, when a high value for the parameter b is introduced, the intermediate stages of the process are far apart to give reasonable results during simulation. As a simple example, consider a logistic process with $b = 3$ starting at $x = 0.10$. The next time period of the process must be 0.27, followed by 0.59 and 0.725. There are very few growth processes that can cover the gap between 10% and 72.5% of the total process in only three time periods. Especially in diffusion and innovation diffusion processes, the development is slower, but eventually the process reaches very high values, and, then, oscillating or even chaotic behaviour appears.

One approach is to use a logistic model with a varying function for the parameter b. This is the method employed in the GRM1 model, where the term $b/(1 - (1 - \sigma)x)$ is replacing the parameter b. Another approach is to select a transformed logistic-like model. This is easily achieved by selecting a modified logistic model of the form:

$$x_{n+1} = bx_n(1 - x_n)^e$$

where $0 < e < 1$. For $e = 1$, this model reduces to the logistic model, whereas, for $e = 0$, it reduces to the exponential model. The fixed point is at $x = 1 - 1/b^{(1/e)}$. This is the minimum level where the first bifurcation starts.

The *modified logistic model* (ML), for $b = 1.3$ and $e = 0.09$, is illustrated in Figure 3.29. The model approaches the equilibrium limit 0.9458. An oscillating form of the modified logistic at $b = 1.3$ and $e = 0.076$ is presented in Figure 3.30 A chaotic process at $b = 1.3$ and $e = 0.073$ for the modified logistic model is presented in Figure 3.31

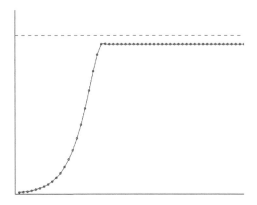

FIGURE 3.29: The modified logistic model (ML)

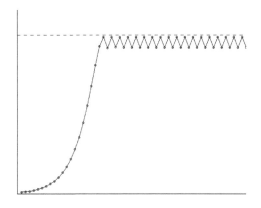

FIGURE 3.30: The modified logistic model (ML) for $b = 1.3$ and $e = 0.076$

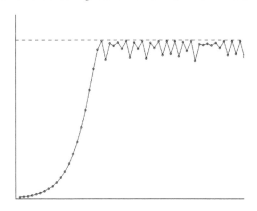

FIGURE 3.31: A chaotic stage of the modified logistic model (ML) for $b = 1.3$ and $e = 0.073$

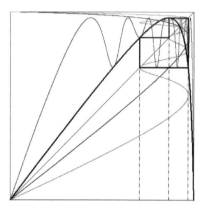

FIGURE 3.32: A third order cycle of the GRM1 model for $b = 1.3$ and $\sigma = 0.025$

Another serious problem when applying the logistic model is that the oscillations following the bifurcation process are quite large for most real life situations. On the other hand, in the cases of the GRM1 and the modified logistic, the amplitude of the oscillations varies according to the values of the parameters selected. A comparative example is given by observing the third order cycle for the logistic, the GRM1 and the modified logistic models (Figure 3.32). The interval of this third order cycle is included inside the chaotic region as illustrated in the bifurcation diagram. By observing this third order cycle, valuable information for the chaotic region can be obtained. During the third order cycle, the logistic model oscillates between 0.1494 and 0.9594, that is, the oscillations cover about 80% of the total amplitude. On the contrary, the GRM1 model is more flexible. The oscillations during the third order cycle can vary according to the values selected for the parameter σ. In Figure 3.32, $\sigma = 0.025$. For this value of σ the GRM1 model shows a third order cycle for $b = 1.3$. The maximum value is 0.9693, and the minimum value is 0.7047. The oscillations therefore cover only 27% of the total amplitude. Accordingly, for lower values of the parameter σ the amplitude of the oscillations during the third order cycle is smaller.

When the parameters for the GRM1 model are $b = 1.2$ and $\sigma = 0.01175$, the higher value of the third order cycle is $x_{\max} = 0.9768$ and the lower value is $x_{\min} = 0.7846$. The third order oscillations cover 19% of the total accepted amplitude of the process. The (x, t) oscillation diagram is presented in Figure 3.33.

During the first time periods the process follows a growth path. Later on, chaotic oscillations appear and the process is stabilised, giving third order chaotic oscillations.

The third order cycle for the modified logistic model is illustrated in Figure 3.34. The parameters selected are $e = 0.0719$ and $b = 1.3$. The minimum value is $x_{min} = 0.8069$ and the maximum value is $x_{max} = 0.9987$. The resulting third order oscillations cover 20% of the total amplitude.

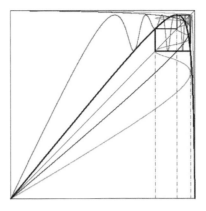

FIGURE 3.33: A third order cycle of the GRM1 model for $b = 1.2$ and $\sigma = 0.01175$

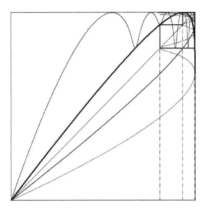

FIGURE 3.34: A third order cycle of the modified logistic model for $b = 1.3$ and $e = 0.0719$

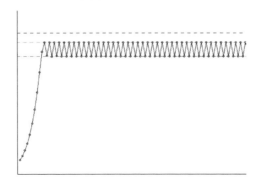

FIGURE 3.35: The first bifurcation cycle of the GRM1 model for $b = 1.3$ and $\sigma = 0.032$

Figure 3.35 illustrates the GRM1 model at the first bifurcation stage. The process ends at a simple oscillation scheme. The parameters are $b = 1.3$ and $\sigma = 0.032$. The maximum and minimum levels of the oscillation are illustrated by the lines with two and one dots respectively. These levels are obtained when $x_{n+2} = x_n = x$. The resulting fourth order equation can be solved explicitly, giving the values for the first bifurcation levels. The fourth order equation for x has, among others, the roots that satisfy the relation

$$x_{n+1} = x_n = x$$

or, after substituting from the GRM1 map

$$x = \frac{bx(1 - x)}{1 - (1 - \sigma)x}$$

or after some simplification:

$$x\left(x - \frac{b - 1}{b - 1 + \sigma}\right) = 0 \tag{3.18}$$

The roots of this equation are $x_1 = 0$ and $x_2 = \frac{b-1}{b-1+\sigma}$. The resulting equation for $x_{n+2} = x_n = x$ can then be transformed to

$$x\left(x - \frac{b - 1}{b - 1 + \sigma}\right) \times$$
$$\times \left[b(b + 1 - \sigma)x^2 - (b + 1 - \sigma)(b + 1)x + (b + 1)\right] = 0 \tag{3.19}$$

Clearly, this fourth order equation has the two roots from (3.18) as well as two other roots:

$$x_{3,4} = \frac{(b + 1) \pm \sqrt{(b + 1)^2 - \frac{4b(b+1)}{b+1-\sigma}}}{2b}$$

It is clear that the equation expressing every higher bifurcation level includes the solutions of the equations expressing the lower bifurcation stages. What is different is the stability of solutions.

Questions and Exercises

1. Show that the fixed point of the map $y = bx(1 - x)$ is $y = x = 1 - \frac{1}{b}$.

2. Compute the derivative of the transformation $y = f(f(x))$, where

$$f(x) = bx(1 - x)$$

at the 2-point orbit given in equation (3.7), and determine from this the range of values of b for which this orbit is attracting.

3. Use the logistic map as a random number generator: For various high values of b, start from an arbitrary value of x, and perform $100,000$ repetitions. Compare the distribution of the $100,000$ values to that of the uniform distribution, as given by the random number generator of your computer. If you found differences, how do you explain these differences?

4. The logistic map is defined on the interval: $0 < x_t \leq 1$. Show that this is indeed the case, *i.e.* that, starting from an x in the interval $[0, 1]$, the process will remain in the interval. Find the related interval for the map $x_{t+1} = 1 + ax_t^2$. For this latter case, draw the bifurcation diagram and find the first and the second bifurcation points.

5. Consider the correct difference equation analogue to the logistic model, as described in page 31. Find the interval where the map is defined and find the formula for the first and second order bifurcation points.

6. Prove that the Gompertz map is defined into the interval $0 < x_t \leq \frac{1}{e}$.

7. Transform the Gompertz and the Mirror Gompertz model to the related discrete maps, and find the interval where every map is defined. Draw the bifurcation diagram for both cases and find the first bifurcation point.

8. Draw the bifurcation diagram for the map $x_{t+1} = b + e^{-ax_t^4}$.

9. Draw the bifurcation diagram for the map $x_{t+1} = b + e^{-ax_t^3}$, and compare the results with the previous case.

Chapter 4

The Delay Logistic Model

4.1 Introduction

Delay models are applied in a large variety of cases in many scientific fields. Consequently, it is difficult to classify the particular models used and to study them systematically. In this section we start with the study of simple delay models, and then proceed to study more complicated models. Special attention is given to the development and presentation of the most frequently used delay models, as well as to related extensions and new delay models of particular interest. Of major importance are the applications of delay models in physiological systems (Glass and Pasternack, 1978; Glass and Mackey, 1988; Glass, 1988, 1991; Glass and Zeng, 1990; de Olivera and Malta, 1987; Cabrera and de la Rubia, 1996; Celka, 1997; Sharkovsky, 1994; Ueda et al., 1994; Bunner, 1999).

Delay models can be divided in two broad categories: The delay models expressed by a difference equation

$$x_{t+m} = f(x_{t+m-1}, x_{t+m-2}, \dots, x_t) \tag{4.1}$$

where m is the delay, and the delay models expressed by a differential equation

$$\dot{x} = f(x_t, x_{t-T}) \tag{4.2}$$

where T is the parameter expressing the delay.

4.2 Delay Difference Models

4.2.1 Simple delay oscillation scheme

Simple oscillation schemes can arise very easily when using delay difference equations. As an example, consider the simple first-degree-polynomial delay form:

$$x_{t+2} = x_{t+1} + ax_t \tag{4.3}$$

When $a = -1$, and for initial values $x_1 = 0.2, x_2 = 0.3$, the resulting two-dimensional map has the hexagonal form illustrated at the top left of Figure 4.1. The hexagon at the top right is formed by considering random initials values (x_1, x_2), uniformly chosen from the interval $[0, 1]$. This iterative scheme generates a discrete oscillation over time (bottom of Figure 4.1).

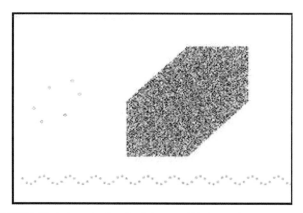

FIGURE 4.1:   A simple delay oscillation scheme ($a = -1$)

A variation of the hexagon in Figure 4.1 can be obtained by introducing a Euclidean distance term in equation (4.1):

$$x_{t+1} = x_t + a\frac{x_{t-1}}{\sqrt{x_t + x_{t-1}}}$$

Figure 4.2 shows the corresponding two-dimensional map for a negative value of a, related to the size of the graph.

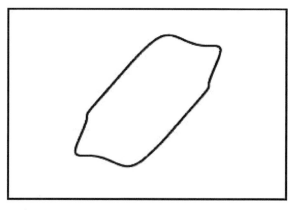

FIGURE 4.2:   Variation of the simple delay model

4.2.2 The delay logistic model

The *delay logistic model* is given by the equation:

$$x_{t+1} = bx_t(1 - x_{t-m}) \tag{4.4}$$

It has several interesting properties, depending of course on the values of b, but, mainly, on the values of the delay parameter m. Clearly, for $m = 0$ equation (4.4) reduces to the logistic equation. The main difference when the delay is present is the formation of a *limit cycle* (Figure 4.3(b)). There is also a marked time period, during which the system takes zero values. The resulting (x_t, x_{t-m}) diagram can be seen at the top of Figure 4.3(b), with the (t, x_t) diagram at the bottom.

The oscillations of the map (4.4) are of a particular form. Each oscillation is followed by a linear term with value of zero. The distance between two successive oscillations depends on the delay parameter: The higher the delay, the higher the distance. The simplest case is illustrated in Figures 4.3(a) and 4.3(b). (The parameters are $m = 1$, $b = 2.271$ and $m = 10$, $b = 1.2865$ respectively).

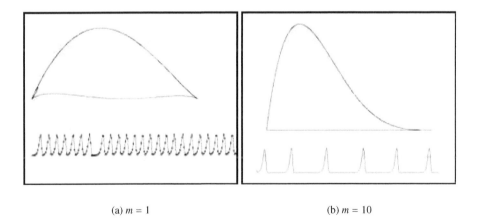

(a) $m = 1$ (b) $m = 10$

FIGURE 4.3: The delay logistic model

The special case with delay $m = 1$ has an interesting property, which can be seen when examining the left corner (corresponding to the origin) of the (x_{t-m}, x_t) diagram closer. The graph exhibits a small "fold," as illustrated in Figure 4.4(a). The $m = 5$ case is even more interesting, with three folds (Figure 4.4(b), $b = 1.44413$).

Table 4.1 summarises the values of b used in our simulations. These are the maximum values for b for which the model makes sense.

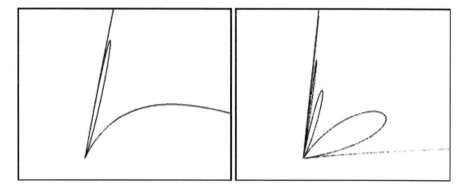

(a) Delay logistic ($m = 1$)   (b) Delay logistic ($m = 5$)

FIGURE 4.4:   Folds in the delay logistic model

TABLE 4.1:   Limiting b values for the delay logistic

m	1	2	3	4	5	$\cdots$	10
b	2.271	1.8395	1.6393	1.52195	1.44413	$\cdots$	1.2865

4.3   Time Delay Differential Equations

The general form of a *time delay differential equation* is given by

$$\dot{x} = f(x_t, x_{t-T}) \tag{4.5}$$

where $T > 0$ is the time delay. The simplest time delay differential equation form is the linear equation

$$\dot{x} = -ax_{t-1} \tag{4.6}$$

The transformation $T = at$ turns (4.6) into

$$\dot{y} = -y_{T-a}$$

where $y_T = x_{T/a}$.

A very interesting observation here is that the parameter a is directly related to the delay; The delay increases along with a. Theory suggests that for $a = \frac{\pi}{2}$ the delay equation has periodic solutions, whereas for $a < \frac{\pi}{2}$ solutions converge to zero, and for $a > \frac{\pi}{2}$, solutions diverge to infinity (Hoppensteadt, 1993).

Using a linear spline approximation method, we can perform a simulation of the above simple time delay equation. The key step in this method is to take the delay interval T and divide it into n equal parts. The iterative process is given by the following scheme:

$$\dot{y}_0 = -ay_n$$
$$\dot{y}_1 = -n(y_1 - y_0)$$
$$\vdots$$
$$\dot{y}_i = -n(y_i - y_{i-1}) \tag{4.7}$$
$$\vdots$$
$$\dot{y}_n = -n(y_n - y_{n-1})$$

This system was solved by using a fourth order Runge-Kutta method of numerical integration, proposed at the end of the nineteenth century by the German mathematicians Carl Runge and Wilhelm Kutta (Runge, 1895; Kutta, 1901). The method is widely used and, when properly applied, gives good results. The selected time increment h must be sufficiently small, and the number of intervals n large enough. Several iterations were tried out for various values of n. When $n = 15$ and $h = 0.001$, the accuracy of the method is good. Figure 4.5 illustrates the two-dimensional Poincaré map,[1] and the periodic solutions of the system when a approaches $\frac{\pi}{2}$. The oscillations are not stable; small changes of the parameter a from the value $\frac{\pi}{2}$ lead to diverging or converging oscillations.

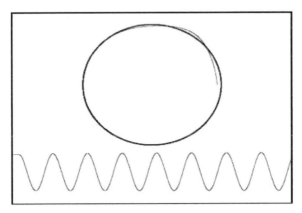

FIGURE 4.5: A simple delay model ($n = 15$)

The behaviour of the simple oscillating scheme simulated above shows that oscil-

---

[1]A Poincaré map is a two-dimensional representation of a multi-dimensional flow of a non-linear system. Geometrically, this map results by cutting the curves of a multidimensional flow by a plane. In our case, a delay equation with delay m is equivalent to an m-dimensional system, and several Poincaré maps may be selected, for instance (x_t, x_{t+m}), (x_{t+m-1}, x_{t+m}), (x_t, x_{t+1}), (x_t, x_{t+2}) and so on. Each of these maps leads to different graphs, showing us different aspects of the flow.

lations appear very naturally in a system with memory. This is because the model expressing the behaviour of the system integrates into the differential-difference equations the past as one or more delay variables. As a consequence, *delay means oscillations.*

Experience confirms the appearance of oscillations in very many systems where delays are inherent. Delays in human decisions on social, economic, technological or political systems may lead to oscillations. Short- or long-term economic cycles, social or political changes, innovation or technological cycles may be expected when time delays are present. Biological cycles, biological clocks and other oscillating mechanisms of humans and other living organisms are some of the numerous cases where delays are present in nature. Chemical oscillations and the Belousov-Zabontiski reactions modelled and studied extensively by Prigogine and his co-workers (Prigogine and Lefever, 1968; Prigogine et al., 1969; Nicolis and Prigogine, 1981; Prigogine, 1995, 1996, 1997) are also examples of the effect of delays in the formation of the intermediate substrates until the formation of the final product.

4.4 A More Complicated Delay Model

Using the same method presented above, we simulate a more complicated delay model, which has a more stable oscillating behaviour. The general delay equation for this model is

$$\dot{x} = -ax_{t-1}\left(1 - x_t^b\right) \tag{4.8}$$

The oscillations of (4.8) are quite stable when $a > \frac{\pi}{2}$, and they exhibit a very interesting behaviour for some values of a. The parameter b is an integer, and two distinct cases are considered, depending on the parity of b. When $b = 1$, the model (4.8) reduces to a continuous delay logistic model.

Of special importance is the case where $b = 2$. When $b = 2$ and $a = 14$, simulation results in the very important form shown in Figure 4.6. The resulting two-dimensional map is an almost perfect *square*, irrespective of the starting point, as long as this point is inside the square, *i.e.* if the starting values are in $[0, 1]$. This rectangular form shows for any $a > \frac{\pi}{2}$, with the sharpness of the corners depending on the value of a. In the case studied in Figure 4.6, the linear spline approximation method was used with $n = 30$, $h = 0.001$ and initial value $x = 0.2$, and it gave very good results. The rectangular oscillations appear to be quite perfect and may be useful in several applications.

The oscillations are quite different when b is odd. The behaviour in this case is similar to that of the discrete delay logistic model (Figure 4.7, $b = 1$, $a = 3.7$, $n = 15$). The delay equation in this case is:

$$\dot{x} = -ax_{t-1}(1 - x_t) \tag{4.9}$$

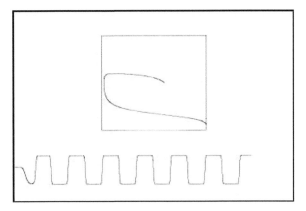

FIGURE 4.6:   Delay model ($b = 2$); a square pattern

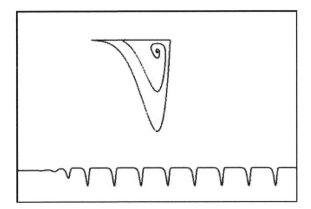

FIGURE 4.7:   Delay logistic model

4.5   A Delay Differential Logistic Analogue

The appropriate differential equation analogue to the delay difference equation (4.9) is:

$$\dot{x}_m = bx(1 - x_m) \tag{4.10}$$

In order to perform a simulation of (4.10), the method of linear spline approximation presented in page 83 will be used. The intermediate step is s and the iteration step is d. The delay term is included only in the first equation. It is the last term to the right of the first equation. By changing this delay term, or replacing it by another, many interesting cases arise.

$$x_1 = x_1 + bx_1(1 - x_m)d$$
$$x_2 = x_2 - \frac{m}{s}(x_2 - x_1)d$$
$$\vdots \tag{4.11}$$
$$x_m = x_m - \frac{m}{s}(x_m - x_{m-1})d$$

Figure 4.8(a) illustrates the (x_1, x_{m+1}) and (t, x_1) diagrams for the case where $b = 1.22$, $d = 0.01$, $s = 3$ and $m = 14$. There is a clear similarity with the previous case of the difference equation above.

The simulation producing Figure 4.8(b) is based on another delay equation:

$$\dot{x}_m = bx_m(1 - x^2) \tag{4.12}$$

The parameters in Figure 4.8(b) are $m = 15$, $s = 6$, $d = 0.001$ and $b = 2.6$. As can be seen from the figure, the (x_1, x_{m+1}) graph is an almost perfect square. Close inspection, however, will show that the corners are not very sharp. This is due to the choice of the parameters and of the integration step. Nevertheless, equation (4.12) is a continuous equation very appropriate for expressing a square. The oscillation turns out to exhibit a step form.

In both cases presented above, b was selected close to its upper limit. However, one can have several intermediate cases of particular interest.

4.6   Other Delay Logistic Models

Another form of a delay logistic model is expressed by the following difference equation:

$$x_{t+1} = ax_t + bx_{t-m}(1 - x_{t-m}) \tag{4.13}$$

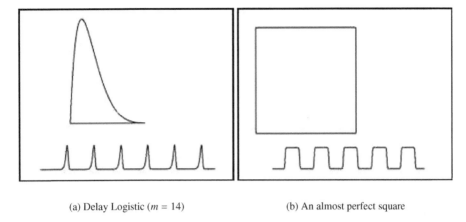

(a) Delay Logistic (*m* = 14) (b) An almost perfect square

FIGURE 4.8: Oscillations and limit cycles in delay logistic models

This is a delay model that integrates the influence of a logistic like interaction to the future. As a simple example, consider the impact of a "word-of-mouth" communication between x adopters and $(1 - x)$ potential adopters of an innovation or an idea, communicated m time units in the past, and influencing people at the present time. This happens very frequently in marketing, economic, social or political systems. The impact of this time-delayed influence to the evolution of x in the future will be studied by varying the width of the delay interval m and the values of the parameters a and b. The ax_t term represents the current adopters who continue, and the $bx_{t-m}(1 - x_{t-m})$ term represents the new adopters.

In this model, limit cycles, bifurcation and chaotic behaviour appear. A large variety of oscillations and chaotic oscillations appear, depending on the level of the delay. But, even in the simplest cases of one or two units of delay, the behaviour of the system is quite complicated. High delay levels ensure chaotic behaviour.

The simplest form of (4.13) is when $m = 1$. The more interesting forms appear in Figure 4.9(a). The six realisations, A through F, illustrate the behaviour of the system when b is 1.706, 1.717, 1.751, 1.758, 1.77 and 1.795 respectively. Throughout, a was set to 0.9 and x_0 was set to 0.1. As b increases, limit cycles are followed by chaotic behaviour.

Figure 4.9(b) examines the case where $m = 2$. Again, limit cycles, bifurcation and then chaos appear as b increases. figures A through F correspond to b values of 1.2, 1.22, 1.241, 1.252, 1.259 and 1.269 respectively ($a = 0.9$ and $x_0 = 0.1$).

The $m = 3$ time delay has an interesting effect on the two-dimensional graph (Figure 4.9(c)). The bifurcation is more complicated, and leads to a chaotic picture with a period doubling. The parameter a and the initial value x_0 are as in the preceding example, whereas b is 0.96, 0.965, 0.99, 0.991, 0.994 and 1.016 respectively.

When $m = 5$, $a = 0.9$ and $x_0 = 0.95$, the time-delay effect is more apparent. (Figure 4.9(d)). The values of b are 0.74, 0.743, 0.747, 0.754, 0.758 and 0.762 respec-

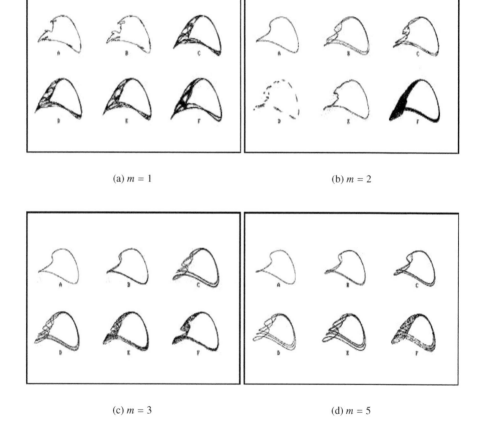

(a) *m* = 1   (b) *m* = 2

(c) *m* = 3   (d) *m* = 5

FIGURE 4.9:   Chaotic attractors of a delay logistic model

tively. The form of the graphs is similar to the $m = 3$ case. Limit cycles, bifurcation and chaotic behaviour are again present.

4.7   Model Behaviour for Large Delays

When the time delay m is large or extremely large, the form of the graphs tends to be stable for changes of the delay parameter and keeps the form of the chaotic attractors. The model is expressed in a high dimensional space, and the form of the attractors is influenced by the delay mechanisms, not merely by the form of the delay equation. This can be seen by comparing the logistic delay model to the Glass model (Mackey and Glass, 1977). The two models look completely different in terms of their analytic equations, but the chaotic attractors and the resulting chaotic oscillations are very similar.

Figure 4.10 illustrates the two-dimensional graphs for the logistic delay model when $m = 40$, $a = 0.9$ and $x_0 = 0.95$. The parameter b here takes the values $0.36, 0.37, 0.38, 0.3813, 0.3818$ and 0.41 respectively.

Graph A represents a classical limit cycle, whereas graph B corresponds to a first order bifurcation, and graphs C and D show second and third order bifurcations respectively. Higher order bifurcations appear in graph E, and chaotic behaviour is present in graph F.

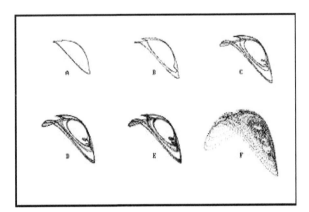

FIGURE 4.10:   Chaotic attractors of the logistic delay model ($m = 40$)

The chaotic behaviour presented in graph F of Figure 4.10 shows great similarities to that arising from a *discrete delay Glass model* of the form

$$x_{t+1} = ax_t + b\frac{x_{t-m}}{1 + x_{t-m}^{10}} \qquad (4.14)$$

The Glass model (graph A) and the logistic model (graph B) are compared in Figure 4.11. The delay parameter is $m = 40$ and $a = 0.9$ for both models, and b is set to 0.18 for the Glass model and 0.4 for the logistic model. The similarities are striking, both in the two-dimensional attractors as well as in the chaotic oscillations, presented in the lower part of the figure.

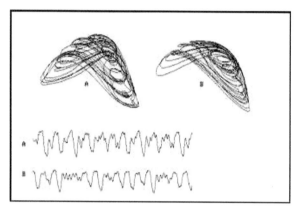

FIGURE 4.11:   Comparing the Glass (*A*) and the logistic (*B*) delay models
$(m = 40)$

Various general models were also tested, with similar results. Such general models are of the form

$$x_{t+1} = ax_t + F(x_{t-m})$$

Examples of specific models that exhibited behaviour similar to that of the Glass and the delay logistic models in the chaotic region are:

LG2	$x_{t+1} = ax_t + bx_{t-m}(1 - x_{t-m}^2)$	
LG3	$x_{t+1} = ax_t + bx_{t-m}(1 - x_{t-m}^3)$	
GRM1	$x_{t+1} = ax_t + b\dfrac{x_{t-m}(1 - x_{t-m})}{1 - (1 - c)x_{t-m}}$	(4.15)
Gauss	$x_{t+1} = ax_t + be^{-c(x_{t-m}-k)^2}$	
Gompertz	$x_{t+1} = ax_t - bx_{t-m}\ln(x_{t-m})$	

We refer to the five models as *delay LG2*, *delay LG3*, *delay GRM1*, *delay Gauss* and *delay Gompertz* respectively. Chaotic behaviour is expressed for the following values of the parameters:

In all these cases, the attractors have the same shape as that illustrated in Figure 4.11.

A different chaotic behaviour is exhibited by the following delay exponential model:

$$x_{t+1} = x_t + bx_{t-m}e^{-cx_{t-m}} \tag{4.16}$$

TABLE 4.2: Parameter values on which various delay models exhibit chaotic behaviour

Model	LG2	LG3	GRM1	Gauss	Gompertz
a	0.9	0.9	0.95	0.9	0.9
b (c, k)	0.3	0.243	0.1 (0.84)	0.17 (4, 0.7)	0.28

Model (4.16) gives a relatively stable limit cycle for various values of the parameters. Figure 4.12 illustrates this limit cycle and the related oscillations when the parameters of the model take the values $a = 0.9$, $b = 3.2$ and $c = 0.8$. A shape like that of a pigeon drawn by a modern painter appears.

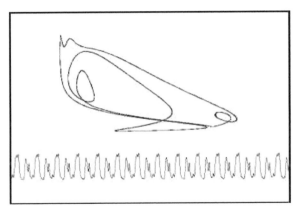

FIGURE 4.12: The delay exponential model

4.8 Another Delay Logistic Model

A somewhat different delay logistic model is given by the following difference equation:

$$x_{t+1} = bx_{t-m} + ax_t(1 - x_t) \tag{4.17}$$

When $m = 1$, this system reduces to a system equivalent to that of the *Hénon* model:

$$x_{t+1} = bx_{t-1} + ax_t(1 - x_t) \tag{4.18}$$

The analysis of system (4.18) when $b = -1$ shows that it has a stable fixed point when

$$x = 1 - \frac{2}{a}$$

which turns into the period-2 orbit with:

$$x = \frac{a + 2 \pm \sqrt{(a-6)(a+2)}}{2a}$$

The Hénon model is usually presented in another form, as a two-dimensional model expressed by a system of difference equations:

$$x_{t+1} = by_t + 1 - ax_t^2$$
$$y_{t+1} = x_t \tag{4.19}$$

However, note that the second equation of (4.19) implies that $y_t = x_{t-1}$. We can therefore substitute this into the first equation, and then the following delay equation results:

$$x_{t+1} = bx_{t-1} + 1 - ax_t^2 \tag{4.20}$$

Equation (4.20) is equivalent to (4.18), as can be easily observed in Figure 4.13(b). The small attractor produces the large attractor by rescaling and translation. The large attractor is the (x_t, x_{t+1}) image produced by equation (4.20) when $a = 1.4$ and $b = 0.3$. The small attractor is the one produced by equation (4.18) when $a = 3.17$ and $b = 0.3$.

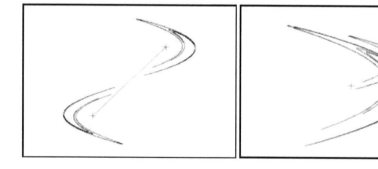

(a) Delay cosine-Hénon   (b) Delay logistic-Hénon

FIGURE 4.13: Delay logistic and delay cosine Hénon variants

In Figure 4.13(a), the two attractors look as if they were produced by reflection and translation of one to the other. The attractor on the top right of Figure 4.13(a) is the Hénon attractor provided by the system (4.19). The cross near this attractor is located at the origin $(0, 0)$, whereas the cross close to the attractor on the left of the same figure is located at $(-\pi, -\pi)$. This attractor results from the system

$$x_{t+1} = by_t + a\cos(x_t)$$
$$y_{t+1} = x_t$$
(4.21)

and its corresponding delay equation

$$x_{t+1} = bx_{t-1} + a\cos(x_t)$$
(4.22)

This system looks very different from (4.19), algebraically, yet the attractors are very similar. The attractor in Figure 4.13(a) is formed when $b = 0.3$ and $a = 3.22$.

Equation (4.22) retains some of the properties of the Hénon map, but it also contains other interesting properties, presented in Figure 4.14 ($a = -0.4$, $b = -1$). The map (4.21) is an area-preserving map, as its Jacobian is $J = b = -1$. The influence of the cosine function in the delay equation is clear: The various forms are repeated many times by following 45° and 135° degrees of symmetry.

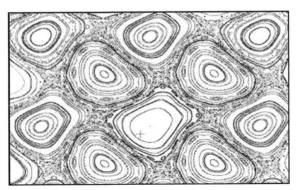

FIGURE 4.14: Chaotic patterns of the delay cosine model

Figure 4.15(a) shows the bifurcation diagram (a, x) for $0 < a < \frac{\pi}{2} + 0.1$ and $b = -1$. The parameter value $a = \frac{\pi}{2}$ gives rise to an interesting carpet-like form (Figure 4.15(b)). An infinite number of *islands* are included inside the *chaotic sea*.

Another carpet-like attractor appears in Figure 4.16(a). Here $b = -1$ and $a = \frac{\pi}{2} - 0.5$. As a is less than the characteristic value $\frac{\pi}{2}$, the islands cover more space inside the chaotic sea. The chaotic sea almost disappears in Figure 4.16(b). The islands have now become very ordered geometric structures, and cover almost the entire plane. Here $a = \frac{\pi}{2} - 1.5$.

There is a systematic way of drawing these graphs. The center of the square-like forms is located at distances proportional to π, whereas the outer dimension of these squares is also equal to π. The squares are also related to each other in terms of their creation over time. With the exception of the central square, all the other squares are generated in groups of four.

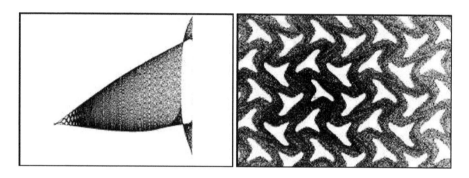

(a) The bifurcation diagram (a, x) (b) A carpet-like map

FIGURE 4.15:   Carpet-like forms

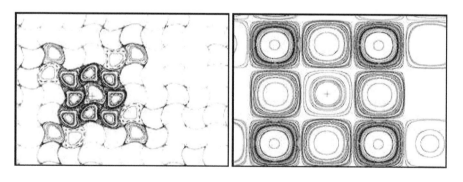

(a) A carpet-like pattern   (b) Groups of square-forms

FIGURE 4.16:   Carpet-like forms

Figure 4.17(a) illustrates the same map as in Figure 4.16(b), but in the latter case only the centers and the corresponding squares are present. When the control parameter is very small, $a = 0.01$, almost perfect squares appear. This is illustrated in Figure 4.17(b). The chaotic sea almost disappears and the non-symmetric islands become geometric objects. The original central square is drawn (starting with an initial value a bit less than $\frac{\pi}{2}$), along with the first group of four surrounding squares (starting with an initial value a bit more than $\frac{\pi}{2}$).

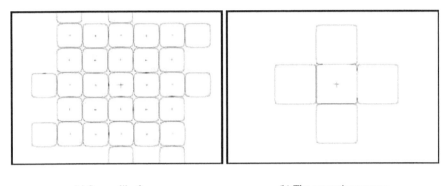

(a) Square-like forms   (b) The generating squares

FIGURE 4.17: Square-like forms

The case of $b = 1$ is illustrated in Figure 4.18(a). The control parameter here is $a = \frac{\pi}{3}$. The building blocks for the construction of the above carpet-like form are two interrelated objects close to the origin (marked by a cross in Figure 4.18(b), where a was set to 0.001). The two forms have the shape of almost perfect squares, with their centers located at $(-\frac{\pi}{2}, \frac{\pi}{2})$ and at $(\frac{\pi}{2}, -\frac{\pi}{2})$.

The bifurcation diagram (b, x) appears in Figure 4.19(a). Period doubling and totally chaotic regions are clearly present. At the end of the diagram the chaotic region has a stochastic character. The oscillations seem to cover the entire plane. Interesting windows and special forms appear when a takes the special values $\frac{\pi}{3}, \frac{\pi}{2}, \frac{2\pi}{3}, \pi$ and $\frac{3\pi}{2}$. The value $\frac{3\pi}{2}$ is the lower limit for the control parameter a. Beyond it, the chaotic sea covers the entire plane, with no islands present. Figure 4.19(b) shows the case where $a = \frac{\pi}{2}$. On the outer part of each pattern we can see 6 small islands.

When $a = \frac{2\pi}{3}$ the fixed point bifurcates into a period-2 orbit. This is illustrated in Figure 4.19(c). The six islands from the preceding case are now separated from the original form, and are inside the chaotic sea. On the boundary one can now see 4 islands. The system is at the starting point of the bifurcation. In Figure 4.19(d) the bifurcation has been completed, and two distinct chaotic objects appear. The control parameter at this point is $a = \pi$.

When the control parameter is increased, a triangular form appears (Figure 4.20(a),

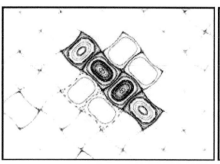
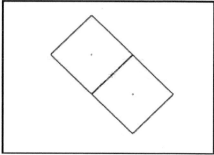

(a) Irregular patterns   (b) The generating squares

FIGURE 4.18:   Irregular patterns

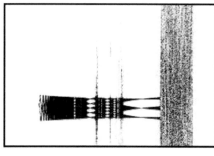
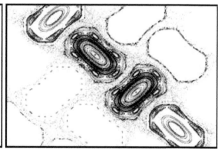

(a) Bifurcation diagram (b, x) (b) Islands in the chaotic sea, $a = \frac{\pi}{2}$

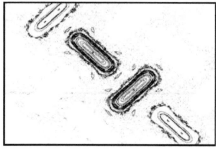
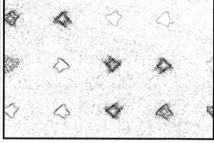

(c) Islands in the chaotic sea, $a = \frac{2\pi}{3}$ (d) The order-2 bifurcation

FIGURE 4.19:   Islands in the chaotic sea

$a = \frac{6\pi}{5} - 0.008$). When a gets closer to $\frac{6\pi}{5}$, the triangular form splits into three objects around the central one (Figure 4.20(b), $a = \frac{6\pi}{5} - 0.02$).

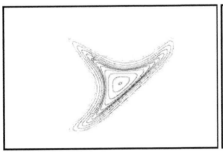   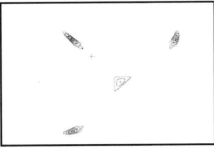

(a) A triangular island   (b) An order-3 bifurcation

FIGURE 4.20:   Triangular islands

The Hénon model was introduced here in order to compare it to the other delay models. However, the Hénon model is a very interesting model, deserving its own study. A more detailed analysis of the Hénon model will be the subject of the next chapter.

Questions and Exercises

1. Determine the sides of the large hexagon appearing on the right of Figure 4.1.

2. Draw the (x_{t-m}, x_{t+1}) and (t, x_{t+1}) graphs for the Gompertz delay model and for various positive values of the parameters a, b and m.

3. Draw the graph of the same model for $m = 40$, $a = 0.9$ and $b = 0.28$, and compare the resulting figure with that of the Glass and the logistic models.

4. Replace the $\cos(x_t)$ term into the delay cosine map (Equation (4.22)) by Taylor approximation to the cosine

$$\cos(x_t) = 1 - \frac{x_t^2}{2}$$

and draw the bifurcation diagram for $b = -1$ and compare the resulting figure with that provided for the delay cosine model. Find the characteristic points and compare them to those of the delay cosine.

5. Repeat the analysis and graphs of the previous problem, but now for the delay map

$$x_{t+1} = -x_{t-1} + a\left(1 - \frac{x_t^2}{c}\right)$$

where c is a positive parameter. Find similarities and differences with the previous case.

6. Find the Jacobian of the map in the previous problem, and generalise your result to a map of the form

$$x_{t+1} = -x_{t-1} + af(x_t)$$

where $f(x_t)$ is a continuous smooth function.

7. Can you draw carpet like forms by using the map resulting from the approximation of the cosine? Justify your answer.

8. What happens to the number of "folds" in Figure 4.4(b) as the delay parameter m increases?

9. Examine what happens when the parameter b in (4.8) is allowed to take non-integer values.

Chapter 5

The Hénon Model

The Hénon model is a two-dimensional model introduced and analysed by Hénon (1976). This model can also be expressed by a delay difference equation:

$$x_{t+1} = bx_{t-1} + 1 - ax_t^2 \qquad (5.1)$$

By introducing a new variable, $y_t = x_{t+1}$, (5.1) is transformed into a system of two ordinary difference equations:

$$
\begin{aligned}
x_{t+1} &= y_t + 1 - ax_t^2 \\
y_{t+1} &= bx_t
\end{aligned}
\qquad (5.2)
$$

The Jacobian determinant of (5.2) is given by:

$$
\det J = \begin{vmatrix} \frac{\partial x_{t+1}}{\partial x_t} & \frac{\partial x_{t+1}}{\partial y_t} \\ \frac{\partial y_{t+1}}{\partial x_t} & \frac{\partial y_{t+1}}{\partial y_t} \end{vmatrix} = -b
$$

When $b = 1$, the resulting map is area preserving, but the system is unstable. The system becomes stable for $0 < b < 1$. A variety of maps appear when a and b vary over a range of values, and the resulting maps take the form of chaotic attractors. We will see a variety of these attractors in this chapter. The most known chaotic attractor of the Hénon model appears when $a = 1.4$ and $b = 0.3$, and is illustrated in Figure 5.1.

5.1 Global Period Doubling Bifurcations in the Hénon Map

Global period doubling bifurcations in the Hénon map have been studied by many researchers, and a very detailed analysis can be found in Murakami et al. (2002). The analysis of the Hénon map is based on (5.1) and (5.2). The fixed points ($x_{t+1} = x_t, y_{t+1} = y_t$) of the map are the solutions of the second-order polynomial:

$$ax^2 - (b - 1)x - 1 = 0$$

namely

$$x_t = \frac{(b - 1) \pm \sqrt{(b - 1)^2 + 4a}}{2a}, \; y_t = x_t$$

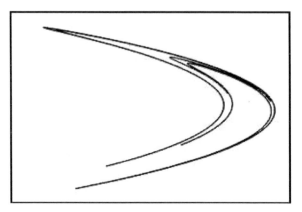

FIGURE 5.1:   The Hénon map

The period-2 orbit can be found from the relations $x_{t+2} = x_t$ and $y_{t+2} = y_t$. The resulting fourth-order equation for x has the two extra solutions

$$x_t = \frac{-(b-1) \pm \sqrt{-3(1-b)^2 + 4a}}{2a}, \; y_t = x_t$$

5.1.1   Period doubling bifurcations when $b = -1$

When $b = -1$, the map has the stable fixed point

$$x_t = \frac{-1 + \sqrt{1+a}}{a}, \; y_t = x_t$$

for $-1 < a < 3$. This turns into the period-2 orbit

$$x_1 = \frac{1 + \sqrt{a-3}}{a}, \; y_1 = \frac{1 - \sqrt{a-3}}{a}$$

$$x_2 = \frac{1 - \sqrt{a-3}}{a}, \; y_2 = \frac{1 + \sqrt{a-3}}{a}$$

(5.3)

at the value $a = 3$. This orbit remains stable for $3 < a < 4$. The second period doubling takes place at $a = 4$, giving rise to the period-4 orbit with

$$x_1 = \frac{1}{\sqrt{a}}, \; y_1 = \frac{1}{\sqrt{a}}\left(1 - \frac{2}{\sqrt{a}}\right)^{1/2}$$

$$x_2 = -y_1, \; y_2 = x_1$$

$$x_3 = x_1, \; y_3 = -y_1$$

$$x_4 = y_1, \; y_4 = x_1$$

(5.4)

The third period doubling bifurcation takes place when $a = 4.12045$, giving rise to a period-8 orbit. This last value of a is the root of:

$$4a(2\sqrt{a} - a) + 1 = 0$$

5.1.2 Period doubling bifurcations when $b = 1$

When $b = 1$, the fixed points of (5.2) are at

$$x = \pm\frac{1}{\sqrt{a}}, y = x \qquad (5.5)$$

These are both unstable when $a > 0$. The period-2 orbit consists of the points

$$x = \pm\frac{1}{\sqrt{a}}, y = -x$$

This orbit becomes unstable when $a = 1$, and we have a bifurcation into a period-4 orbit. The period-4 orbit undergoes a period doubling bifurcation at $a = \frac{8-\sqrt{12}}{4} = 1.13397$ and leads to a period-8 orbit. The period-8 orbit bifurcates into the period-16 orbit at $a = 1.15135$. We leave the details to the reader.

5.2 The Cosine-Hénon Model

During the simulation process related to the global period doubling bifurcation, difficulties arise on the estimation of the starting parameters, as well as of the local or global instabilities related to the period doubling. It is, therefore, more convenient to use an analogue of the Hénon system, based on trigonometric functions:

$$\begin{aligned}
x_{t+1} &= by_t + 2a\left(\cos\left(x_t\right) - 1\right) + 1 \\
y_{t+1} &= x_t
\end{aligned} \qquad (5.6)$$

This model is called the *cosine Hénon model.* The relation between the two models is that the Hénon model (5.2) is the model obtained from (5.6) when the cosine term is replaced by its second-order Taylor approximation:

$$\cos(x_t) = 1 - \frac{x_t^2}{2}$$

Therefore, this new model retains many of the properties of the Hénon model, but it also has other distinct properties. This model accepts a wider range of initial values in the (x, y) plane. Furthermore, the values of the control parameter related to the period doubling bifurcations of the Hénon map are close to the corresponding values of the control parameter for the cosine model ($b = -1$). The fixed point at the critical value $a = 3$ is illustrated in Figure 5.2(a). Also present in this case are 7 islands on the boundary. Figure 5.2(b) illustrates the period-2 orbit at $a = 3.23$. The system is in sub-critical state, just before the completion of the next global period doubling. We can see 11 islands on the boundary.

Figure 5.2(c) illustrates the period-4 orbit of the cosine map with $b = -1$ and $a = 4.11$.

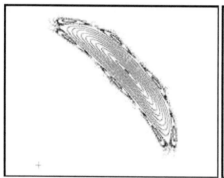

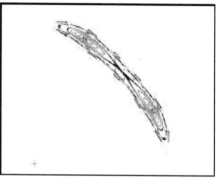

(a) The fixed point ($a = 3$)   (b) The period-2 orbit

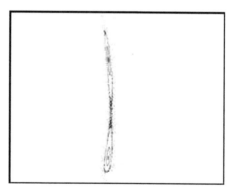

(c) The period-4 orbit

FIGURE 5.2:   Orbits in the cosine Hénon model

5.3   An Example of Bifurcation and Period Doubling

Figures 5.3(a) and 5.3(b) illustrate some interesting cases regarding the bifurcation and period doubling when the parameter b of the cosine Hénon model takes the value $b = 1$. The system will undergo period doubling when $a = 1$. Figures 5.3(a), 5.3(b), 5.4(a) and 5.4(b) show the map when the control parameter is $a = 0.8$, 0.98, 1.02 and 1.3 respectively. In Figure 5.3(a), a double structure appears. After the period doubling, we see that, in each of the original structures, the central island has turned into three separate forms. In Figure 5.4(b), the period-4 islands are completely separated.

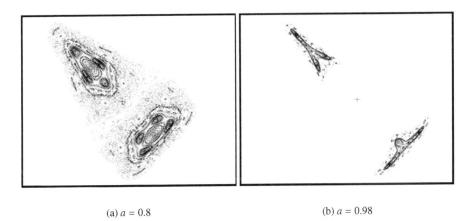

(a) $a = 0.8$ (b) $a = 0.98$

FIGURE 5.3:   Bifurcation forms in the cosine Hénon model

5.4   A Differential Equation Analogue

A differential equation analogue to the Hénon model is given by the following system of two coupled differential equations

$$\dot{x} = by + 1 - ax^2 - x$$
$$\dot{y} = x - y$$

(5.7)

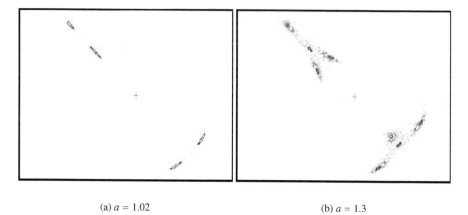

<table>
<tr><td>(a) $a = 1.02$</td><td>(b) $a = 1.3$</td></tr>
</table>

FIGURE 5.4:   Higher-order bifurcations in the cosine Hénon model

This system has fixed points at

$$x = y = \frac{b - 1 \pm \sqrt{(b - 1)^2 + 4a}}{2a}$$

When $b = -1$, these fixed points become:

$$x = y = \frac{-1 \pm \sqrt{1 + a}}{a}$$

which is precisely the same as the fixed points for (5.2).

Similarly, when $b = 1$, the provided value is $x = y = \pm \frac{1}{\sqrt{a}}$ which is also the same value as that provided from the difference equation case.

5.5   Variants of the Hénon Delay Difference Equation

A number of variants of the Hénon model give rise to some interesting attractors. We discuss a couple of these variants, involving either higher-order delays (sections 5.5.1, 5.5.2), or simply changes in the way the first-order terms are introduced (sections 5.5.3, 5.5.4).

5.5.1   The third-order delay model

An extension to higher delay dimensions is given by

$$x_{t+1} = bx_{t-3} + ax_t(1 - x_t) \tag{5.8}$$

Similar to (5.2), the difference delay equation (5.8) can be turned into a system of four ordinary difference equations. The number of time delays in the delay difference equation is equal to the number of new equations of the ordinary difference equation system. The model (5.8) exhibits three time-delay periods, and is characterised by a very interesting attractor when the parameters are $a = 3.2$ and $b = 0.2$ (Figure 5.5).

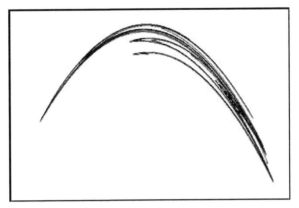

FIGURE 5.5: A third-order delay model

5.5.2 Second-order delay models

The following variant, with a second-order delay term, is extremely simple, having only one parameter b:

$$x_{t+1} = -bx_t + x_t x_{t-2} \tag{5.9}$$

The attractors of the model are however very interesting. Figure 5.6 for instance shows the attractor when $b = 1.37$.

Another interesting variant with a second-order delay is the following:

$$x_{t+1} = x_t + a(x_{t-2} - x_{t-2}^3) \tag{5.10}$$

When $a = 0.52$, the resulting map and the related oscillations (t, x) are illustrated in Figure 5.7.

A more complicated model is given by:

$$x_{t+1} = bx_{t-1}^2 - ax_t + x_t x_{t-2} \tag{5.11}$$

This is a two-parameter delay model with both first- and second-order delay terms. There is a characteristic limit cycle obtained when the parameters have values $a = 1.2$ and $b = 0.75$. The repeated oscillations and the limit cycle are presented in Figure 5.8.

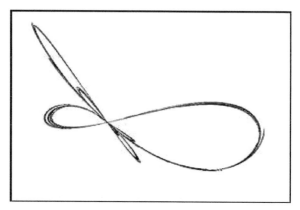

FIGURE 5.6:   A second-order delay model

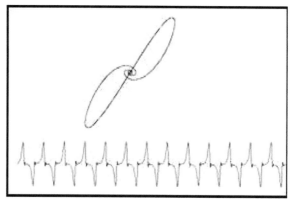

FIGURE 5.7:   The order-2 delay model

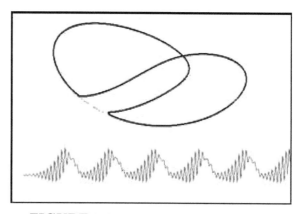

FIGURE 5.8:   A complicated delay model

5.5.3 First-order delay variants

Interesting first-order delay variants of the Hénon model arise when one of the two terms on the right-hand side of (5.1) is altered.

For example, if we replace the quadratic term with a delay logistic term, the resulting model has the form:

$$x_{t+1} = bx_{t-1} + ax_t(1 - x_{t-1}) \qquad (5.12)$$

Figure 5.9 illustrates the two-dimensional map, and the related oscillations, when $a = 2.35$ and $b = 0.2$.

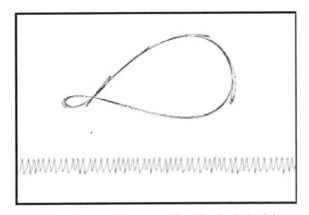

FIGURE 5.9: A Hénon model with a logistic delay term

Another variant arises by altering the first term:

$$x_{t+1} = bx_t x_{t-1} + 1 - ax_t^2 \qquad (5.13)$$

For $a = 1.72$ and $b = 0.6$, the resulting map and related oscillations are illustrated in Figure 5.10.

As in the case of the original Hénon model, it is more convenient, for further analysis, to transform (5.13) into a system of two difference equations:

$$\begin{aligned} x_{t+1} &= x_t y_t + 1 - ax_t^2 \\ y_{t+1} &= bx_t \end{aligned} \qquad (5.14)$$

The Jacobian determinant of (5.14) is

$$\det J = -bx_t.$$

This indicates that the model (5.14) has properties different from those of the Hénon model.

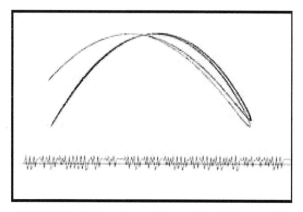

FIGURE 5.10:   A variation of the Hénon model

5.5.4   Exponential variants

The introduction of an exponential term into (4.20) gives rise to a variant with a very complex attractor (Figure 5.11, $a = 0.8$, $b = 0.36$), which we will call the *Henia attractor*. The equation is

$$x_{t+1} = bx_{t-1} + 1 - ax_t^2 e^{-ax_{t-1}} \tag{5.15}$$

FIGURE 5.11:   An exponential variant of the Hénon model

5.6 Variants of the Hénon System Equations

Other interesting variants arise from altering the system (5.2). An example is the following:

$$x_{t+1} = bx_t + cy_t - 2x_ty_t$$
$$y_{t+1} = a(y_t - x_t) \tag{5.16}$$

For appropriate values of the parameters, interesting stone-shaped attractors appear (Figure 5.12 ($a = 0.8$, $b = 0.48$, $c = 1.1$)).

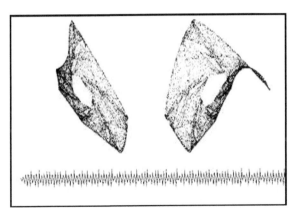

FIGURE 5.12:   A 2-difference equations system

Another set of extensions involves the introduction of a third variable:

$$x_{t+1} = bx_t - y_t - x_tz_t$$
$$y_{t+1} = -cz_t + x_tz_t \tag{5.17}$$
$$z_{t+1} = a(z_t - x_t)$$

Depending on the values of the three parameters, this model exhibits limit cycles, bifurcation and finally chaotic behaviour. Figure 5.13 illustrates these cases. The parameters a and c are fixed at 0.95 and 0.4 respectively, while b takes the values 0.62, 0.628, 0.6337, 0.6348, 0.6355 and 0.637 respectively. The graphs represent the (x, y) map.

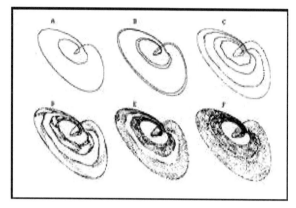

FIGURE 5.13:   A three-dimensional model

5.7   The Holmes and Sine Delay Models

5.7.1   The Holmes model

The Holmes model can be thought of as an extension of the Hénon model (Holmes, 1979a). The general form of the Holmes model is:

$$x_{t+1} = bx_{t-1} + ax_t - x_t^3$$

The Holmes model, when $b = -0.2$ and $a = 2.765$, provides a characteristic sigmoid map and quite complicated chaotic oscillations (Figure 5.14). When $b > 0$, the

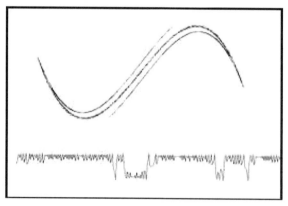

FIGURE 5.14:   The Holmes model ($a = 2.765, b = -0.2$)

Holmes model gives a rich sigmoid form (Figure 5.15, $b = 0.3$ and $a = 2.4$).

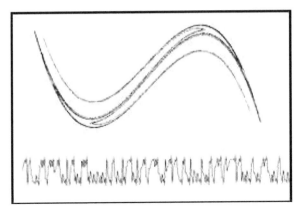

FIGURE 5.15: The Holmes model ($a = 2.4, b = 0.3$)

5.7.2 The sine delay model

The model proposed by Holmes is an approximation[1] of a sinusoidal one, called the *sine delay model*, of the form:

$$x_{t+1} = bx_{t-1} + a\sin(x_t) \tag{5.18}$$

This model, for relatively small values of a, shows similar behaviour to that of the Holmes model. However, when a takes large values, the map changes into a series of sigmoid forms. The number of maxima or minima of the resulting curves can be estimated according to the number of solutions of the equation

$$(1 - b)x - a\sin(x) = 0.$$

In the example presented in Figure 5.16, the delay parameter is fixed at $b = 0.3$ in both cases, and the parameter a is set to $a = 3.0772$ for the case A and $a = 2 \times 3.0772$ for the case B. It is clear, by observing the resulting maps, that by doubling the value of a, the number of maxima or minima is also doubled.

When b is small, there is a close relation between the maxima and minima of this map, and the maxima and minima of the function

$$y = f(x) = (1 - b)x - a\sin(x).$$

The relation can be seen in Figure 5.17(a). Note further that the range of x values for the map ends exactly at the last solution to $f(x)$ (the solutions to $f(x) = 0$ are exactly the fixed points of the sine delay map). Figure 5.17(b) shows the case where $b = 0.3$ and $a = 6.5(1 - b)\pi$.

When b takes values closer to 1, say $b = 0.88$, the situation changes radically, as the attractor is now separated in two very similar parts (Figure 5.18(a)). There is now

---

[1]The approximation can be seen by replacing $\sin(x)$ with its third degree Taylor approximation, $x - \frac{1}{6}x^3$.

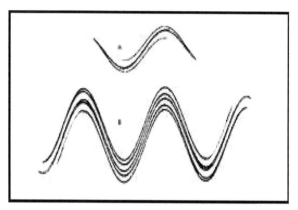

FIGURE 5.16:   The sine delay map ($b = 0.3$)

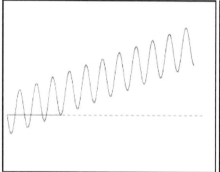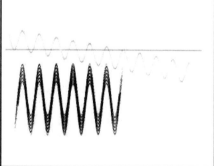

(a) Solutions to the fixed point equation for the sine delay model

(b) Maxima and minima of the sine delay model and the corresponding fixed point equation

FIGURE 5.17:   The sine delay model

little relation with the maxima and minima of the fixed point equation. The values $a = 3$ and $b = 0.8$ produce the very interesting attractor of Figure 5.18(b).

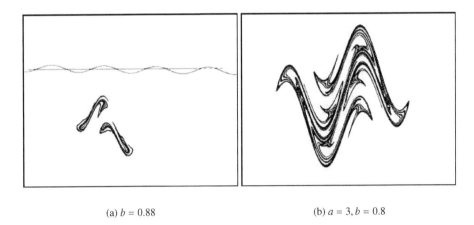

(a) $b = 0.88$ (b) $a = 3, b = 0.8$

FIGURE 5.18: The sine delay model for large b

Things change quite a bit when $b = 1$. In this case, the map takes the form

$$x_{t+1} = x_{t-1} + a \sin(x_t).$$

The resulting simulation is shown in Figure 5.19(a) for $a = 1.2$. The process follows a random sequence that gradually occupies the entire (x_{t-1}, x_t) plane. The basic forms of the two generating shapes can be seen more clearly in Figure 5.19(b), where $a = 0.8$.

Both shapes are identical but they are a ninety-degrees rotation of each other. Both shapes become closer to a square with rounded corners as the parameter a takes very small positive values (Figure 5.20, $a = 0.1$).

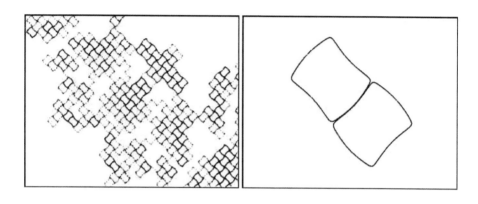

(a) $a = 1.2$ (b) Close look at the two shapes, $a = 0.8$

FIGURE 5.19:   The sine delay model for $b = 1$

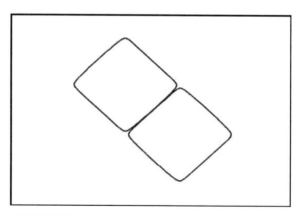

FIGURE 5.20:   The sine delay model ($a = 0.1$)

Questions and Exercises

1. Verify that the fixed points and period-doubling bifurcations of the Hénon map for $b = 1$ and $b = -1$ are as described in section 5.1.

2. Examine the behaviour of the Hénon map for $b = 3$.

3. For each of the maps (5.9), (5.10) and (5.11):

   (a) Determine the fixed points and second-order fixed orbits, and their stability.

   (b) Rewrite them as systems of 3 difference equations without delay, and calculate the Jacobian determinants of these systems.

4. Determine the fixed points of the sine delay model given by equation (5.18) when $b = 1$, and their stability. Use this information to explain the carpet-like forms arising when $b = 1$ (Figure 5.19(a)).

5. Continuing from the previous question, run simulations with different parameter values, and examine whether the carpet-like formations are present, and how they spread across the plane.

6. Explain why for small b there is a relationship between the maxima and minima of the sine delay model (x_t, x_{t+1}) map and the maxima and minima of the equation for the fixed points of the map.

7. Estimate the number of maxima and minima of the fixed point equation for the sine delay model:
$$f(x) = (1 - b)x - a \sin(x)$$

Hint: Find an equation for those maxima and minima from $f'(x) = 0$, and estimate the n beyond which $f(c) > 0$ at the minima.

Chapter 6

Three-Dimensional and Higher-Dimensional Models

In this chapter, we present a number of higher-dimensional models that exhibit chaotic behaviour. Before we delve into particular examples, though, a general discussion of the behaviour of such systems near their equilibrium points is necessary.

6.1 Equilibrium Points and Characteristic Matrices

Let us consider a general autonomous system of the form

$$\dot{x} = \dot{x}(x, y, z)$$
$$\dot{y} = \dot{y}(x, y, z) \qquad (6.1)$$
$$\dot{x} = \dot{z}(x, y, z)$$

and suppose $\mathbf{x}_0 = (x_0, y_0, z_0)$ is an equilibrium point for it, *i.e.* a point where

$$\dot{\mathbf{x}}_0 = (\dot{x}_0, \dot{y}_0, \dot{z}_0) = (0, 0, 0).$$

Consider the matrix:

$$\mathbf{A} = \begin{bmatrix} \frac{\partial \dot{x}}{\partial x} & \frac{\partial \dot{x}}{\partial y} & \frac{\partial \dot{x}}{\partial z} \\ \frac{\partial \dot{y}}{\partial x} & \frac{\partial \dot{y}}{\partial y} & \frac{\partial \dot{y}}{\partial z} \\ \frac{\partial \dot{z}}{\partial x} & \frac{\partial \dot{z}}{\partial y} & \frac{\partial \dot{z}}{\partial z} \end{bmatrix}$$

Then, at a point $\mathbf{x} = \mathbf{x}_0 + \mathbf{h}$ sufficiently close to $\mathbf{x}_0$, we can assume, using a first order Taylor approximation, that $\dot{\mathbf{x}}$ is given by:

$$\dot{\mathbf{x}} = \dot{\mathbf{x}}_0 + A \cdot \mathbf{h} = A \cdot \mathbf{h}$$

Suppose now that the matrix A has a real eigenvalue, say λ_1, with eigenvector $\mathbf{h}_1$. Then, the orbit of a particle placed at $\mathbf{x}_0 + \delta \mathbf{h}_1$ will move away from $\mathbf{x}_0$, in the direction of the vector $\mathbf{h}_1$, since its position after a short time t would be:

$$\mathbf{x}(t) \approx (\mathbf{x}_0 + \delta \mathbf{h}_1) + \dot{\mathbf{x}} \cdot t = \mathbf{x}_0 + (\delta + \lambda_1 t)\mathbf{h}_1.$$

So any equilibrium point with a positive eigenvalue will be unstable. To determine these eigenvalues, we need to compute the roots of the *characteristic polynomial*

$$|\mathbf{A} - \lambda\mathbf{I}| = \begin{vmatrix} \frac{\partial \dot{x}}{\partial x} - \lambda & \frac{\partial \dot{x}}{\partial y} & \frac{\partial \dot{x}}{\partial z} \\ \frac{\partial \dot{y}}{\partial x} & \frac{\partial \dot{y}}{\partial y} - \lambda & \frac{\partial \dot{y}}{\partial z} \\ \frac{\partial \dot{z}}{\partial x} & \frac{\partial \dot{z}}{\partial y} & \frac{\partial \dot{z}}{\partial z} - \lambda \end{vmatrix}$$

The matrix $\mathbf{A} - \lambda\mathbf{I}$ is called the *characteristic matrix* of the system.

6.2   The Lotka-Volterra Model

As an illustration of the above theory in a simpler case, let us consider a classical two-dimensional system, the *Lotka-Volterra* model, also known as the *predator-prey* model. In its simplified form, the Lotka-Volterra model has the form:

$$\dot{x} = x - xy$$
$$\dot{y} = -y + xy \tag{6.2}$$

In the terminology of Chapter 7, this is a conservative system with first integral of motion:

$$f(x, y) = xye^{-x-y}$$

It has two equilibrium points, at $(x, y) = (0, 0)$ and at $(x, y) = (1, 1)$. The characteristic polynomial is given by

$$\begin{vmatrix} -y + 1 - \lambda & -x \\ y & x - 1 - \lambda \end{vmatrix} = \lambda^2 - (x - y)\lambda + (x + y - 1)$$

At $(x, y) = (0, 0)$, the characteristic equation is

$$\lambda^2 - 1 = 0$$

and the system has two real eigenvalues $\lambda_{1,2} = \pm 1$. The eigenvectors correspond to the two directions where there are no predators, and where there are no prey. Since one of these eigenvalues is positive, the system is unstable. At $(x, y) = (1, 1)$, the characteristic equation is

$$\lambda^2 + 1 = 0$$

and the system has two imaginary eigenvalues, $\lambda_{1,2} = \pm i$. The system will cycle around this point.

Using linearization, we see that near $(0, 0)$, the system can be replaced by the simpler system

$$\dot{x} = \quad x$$
$$\dot{y} = -y$$

whose orbits are the hyperbolas $xy = c$. Near $(1, 1)$, if we use coordinates $(\xi, \eta) = (x - 1, y - 1)$ centred at 1, 1, the system can be replaced by the simpler system

$$\dot{\xi} = -\eta$$
$$\dot{\eta} = \xi$$

whose orbits are counterclockwise circles $\xi^2 + \eta^2 = (x - 1)^2 + (y - 1)^2 = c$.

6.3 The Arneodo Model

The simplest three-dimensional chaotic models are based on a system of three linear differential equations, on which we add a non-linear term. An example of such a system is the *Arneodo model*:

$$\dot{x} = y$$
$$\dot{y} = z \tag{6.3}$$
$$\dot{z} = -\left(z + sy - mx + x^2\right)$$

where s and m are parameters.

This model was proposed by Arneodo et al. (1979, 1980, 1981a), in their work on systems that exhibit multiple periodicity and chaos. The Jacobian determinant of the model is

$$J = -m + 2x.$$

There are two stationary (equilibrium) points, the origin $(0, 0, 0)$, and $(m, 0, 0)$.

In Figure 6.1, the (x, y) diagram of the system is shown ($s = 3.8$ and $m = 7.5$, the initial values are ($x_0 = 0.5$ and $y_0 = z_0 = 1$). An integration method was used with integration step $d = 0.001$.

A slight variation, replacing the last equation in (6.3) with

$$\dot{z} = -\left(z + sy - mx + x^3\right)$$

leads to a model with a more stable behaviour. Figure 6.2 shows the projections of the three-dimensional solution path of this system to the three planes (x, y), (x, z) and (y, z) respectively. The parameters are as above, except that $m = 5$. The bottom part of this figure shows the chaotic oscillations in the (y, t) space.

The resulting paths of this modified system have a somewhat more complicated form than those of (6.3). There are three stationary points:

$$(\quad 0, 0, 0)$$
$$(\quad \sqrt{m}, 0, 0) \tag{6.4}$$
$$(- \sqrt{m}, 0, 0)$$

The Jacobian of the modified system is $J = -m + 3x^2$.

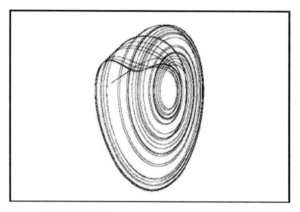

FIGURE 6.1:   The Arneodo model

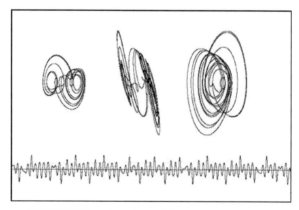

FIGURE 6.2:   A modified Arneodo model

6.4   An Autocatalytic Attractor

Autocatalytic reactions often give rise to chaotic phenomena. The representation of chemical reaction models by mathematical non-linear models is possible and has been studied extensively (Prigogine and Lefever, 1968; Prigogine et al., 1969; Nicolis, 1971; Nicolis and Portnow, 1973; Nicolis and Prigogine, 1977, 1981; Nicolis et al., 1983). It is more difficult to find a good representation of the numerous intermediate stages of a chemical reaction. The number of these stages influences the complexity of the system, and, consequently, gives rise to a process with the chaotic behaviour. Non-linear modelling theory suggests that at least one non-linear part must be present in each reaction cycle in order for chaotic phenomena to appear.

A three-dimensional chaotic model is perhaps the simplest case, representing an autocatalytic reaction with three stages. Such a model, for appropriate parameter values, may demonstrate chaotic behaviour, giving rise to a three-dimensional chaotic attractor. A typical such model is given by the following system:

$$\dot{x} = k + \lambda z - xy^2 - x$$
$$\dot{y} = \frac{xy^2 + x - y}{e} \tag{6.5}$$
$$\dot{z} = y - z$$

When the parameters of system (6.5) take the values $k = 0.7567$, $\lambda = 0.3$ and $e = 0.013$, Figure 6.3 arises. On the right side we see the two-dimensional (x, y) phase portrait, while on the left side we see the temporal variation of chaotic oscillations in the y direction (*i.e.* the (t, y) diagram).

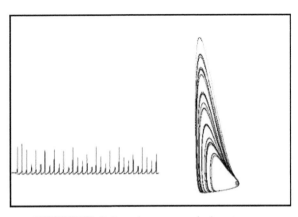

FIGURE 6.3:   An autocatalytic attractor

A three-dimensional view (x, z, y) of the phase portrait of the autocatalytic reaction is presented in Figure 6.4(a). The view from the (y, z, x) coordinate system is presented in Figure 6.4(b). The parameters in both cases are $k = 2.5$, $m = 0.017$ and $e = 0.013$, and the integration step is $d = 0.0001$. In this case the system (6.5) has been modified slightly, with the first equation replaced by:

$$\dot{x} = \left(\frac{1}{1 + k} + m\right)(k + z) - xy^2 - x$$

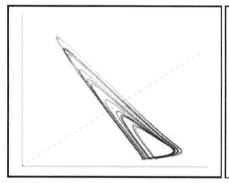
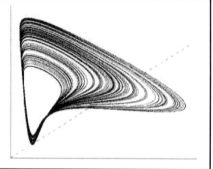

(a) In the (x, z, y) coordinate system   (b) In the (y, z, x) coordinate system

FIGURE 6.4:   Three-dimensional views of the autocatalytic model

6.5   A Four-Dimensional Autocatalytic Attractor

The following four-dimensional model expresses a sequence of four chemical autocatalytic reactions:

$$\dot{x}_1 = x_1 - x_1 x_2 - 0.24 x_1^2 + 200 x_4 + a$$
$$\dot{x}_2 = x_1 x_2 - x_2 - 10 x_2 x_3$$
$$\dot{x}_3 = 0.01 + 10 x_2 x_3 - 20 x_3 \qquad (6.6)$$
$$\dot{x}_4 = 0.24 x_1^2 - 100 x_4$$

The non-linear character of (6.6) is due to the second order term in x_1, which appears in the first and fourth equations, and to the couplings $x_1 x_2$ in the first and the second equations and $x_2 x_3$ in the second and third equations. The control parameter, a, has value -0.12, and the integration step is $d = 0.003$.

Figure 6.5(a) illustrates, from top to bottom, the (x_1, t), (x_2, t), (x_3, t) and (x_4, t) diagrams. The chaotic behaviour is similar, yet more complicated than in the previous cases, whereas the time oscillations are of a purely chaotic nature, with several peaks and intermediate chaotic stages. This is a very good example of chemical oscillations and the related chaotic waves. Figure 6.5(b) shows the three-dimensional (x_1, x_2, x_3) diagram representing the chaotic attractor of (6.6).

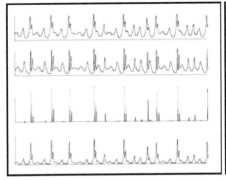
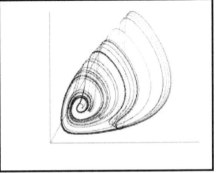

(a) Temporal chaotic oscillations   (b) The (x_1, x_2, x_3) diagram

FIGURE 6.5:   A four-dimensional autocatalytic model

6.6   The Rössler Model

One of the simplest three-dimensional models is the *Rössler model* (see Rössler, 1976d):

$$
\begin{aligned}
\dot{x} &= -y - z \\
\dot{y} &= \;\;\; x - ez \\
\dot{z} &= f \;\; + xz - mz
\end{aligned}
\tag{6.7}
$$

The model contains a single non-linear term, xz, which enters in the third equation, and three parameters, e, f and m.

Figure 6.6(a) is a three-dimensional view of the Rössler model when the parameters are: $e = 0.2$, $f = 0.2$ and $m = 5.7$. Figure 6.6(b) shows, from left to right, the (x, y), (x, z) and (y, z) views, along with the temporal variation in the x, y and z coordinates respectively.

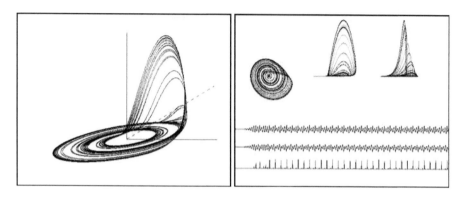

(a) A three-dimensional view of the Rössler attractor

(b) Chaotic oscillations and two-dimensional views

FIGURE 6.6:   The Rössler attractor

6.6.1   A variant of the Rössler model

An interesting variant of the Rössler model is given by the following system of differential equations:

$$\dot{x} = -y - z$$
$$\dot{y} = \quad x + ay \tag{6.8}$$
$$\dot{z} = bx + xz - cz$$

where a, b and c are the parameters of the model.

In all the cases that follow, the parameters b and c are specified in terms of a as follows:

$$b = a$$
$$c = 1 + \frac{1}{a} \tag{6.9}$$

In Figure 6.7(a), a is set to 0.2, and the first order limit cycle of system (6.8) appears.

The order two limit cycle is presented in Figure 6.7(b), where $a = 0.26$. Figure 6.7(c) illustrates the first order cycle when $a = 0.35$, whereas in Figure 6.7(d) we see the second order cycle ($a = 0.42$).

When $a = 0.5$, total chaos ensues. This is illustrated in Figure 6.8(a). This chaotic form is usually referred to as *spiral chaos*, after the characteristic spiral paths of the resulting chaotic image.

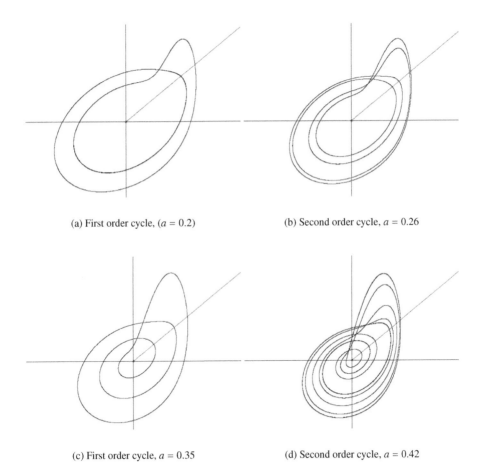

(a) First order cycle, $(a = 0.2)$　　　　　(b) Second order cycle, $a = 0.26$

(c) First order cycle, $a = 0.35$　　　　　(d) Second order cycle, $a = 0.42$

FIGURE 6.7:　Limit cycles in a variant of the Rössler model

A more complicated case of spiral chaos appears when $a = 0.6$. Now the trajectories pass close to the origin. A three-dimensional view of the total spiral chaos is
illustrated in Figure 6.8(b). Figure 6.9 shows, from left to right, the (x, y), (y, z) and
(x, z) views for the same value of a.

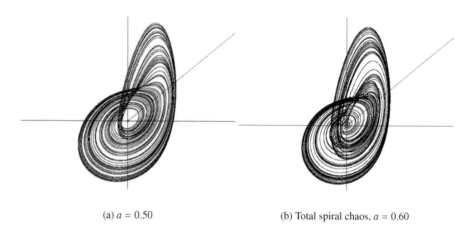

(a) $a = 0.50$ (b) Total spiral chaos, $a = 0.60$

FIGURE 6.8:   Spiral chaos in the variant of the Rössler model

The characteristic matrix of the Rössler model is

$$\mathbf{A} - \lambda\mathbf{I} = \begin{bmatrix} -\lambda & -1 & -1 \\ 1 & a - \lambda & 0 \\ b + z & 0 & x - c - \lambda \end{bmatrix}$$

The corresponding characteristic polynomial is

$$|\mathbf{A} - \lambda\mathbf{I}| = \begin{vmatrix} -\lambda & -1 & -1 \\ 1 & a - \lambda & 0 \\ b + z & 0 & x - c - \lambda \end{vmatrix}$$

$$= -\lambda^3 + (x + a - c)\lambda^2 - (ax - ac + z + b + 1)\lambda + (x + az + ab - c)$$

This polynomial determines the stability of the system at its equilibrium points. In
our case, the equilibrium points are the solutions of the system:

$$y + z = 0$$
$$x + ay = 0 \tag{6.10}$$
$$bx - cz + xz = 0$$

The first equilibrium point is at the origin, $M_1 = (0, 0, 0)$. The most interesting
paths are in the neighbourhood of this point. The other equilibrium point is located

FIGURE 6.9: Plane views of the total spiral chaos in the variant of the Rössler model ($a = 0.60$)

at:

$$M_2 = (x, y, z) = \left(c - ab, -\frac{c - ab}{a}, \frac{c - ab}{a}\right) \tag{6.11}$$

The characteristic equation at the origin is

$$\lambda^3 + (c - a)\lambda^2 + (1 + b - ac)\lambda + (c - ab) = 0$$

The characteristic equation at M_2 is

$$\lambda^3 - a(1 - b)\lambda^2 + \left(1 - a^2b + \frac{c}{a}\right)\lambda - (c - ab) = 0 \tag{6.12}$$

If a, b, c are such that $c - ab > 0$, then equation (6.12) has at least one positive root. This in turn makes M_2 into an unstable point.

The variant of the Rössler model is illustrated in Figure 6.10. The (x, y) graph is presented, where the parameters are $a = 0.4$, $b = 0.3$ and $c = 4.8$. In the same figure, the characteristic polynomial is illustrated. It is a third-degree polynomial with only one real and two complex roots (eigenvalues).

6.6.2   Introducing rotation into the Rössler model

A simple way to introduce rotation into the Rössler model is to insert a simple rotation in two of the equations of the model. The resulting equations have the form:

$$\dot{x} = -z \sin(h) - y \cos(h)$$
$$\dot{y} = x \cos(h) + ey \sin(h) \tag{6.13}$$
$$\dot{z} = f + xz - mz$$

The rotation angle h is a function of the distance r from the origin. For the following simulations, we used

$$h = 0.27 \sqrt{x^2 + y^2}.$$

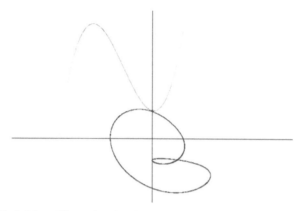

FIGURE 6.10:   View of a simple cycle of the variant of the Rössler model
($a = 0.4$, $b = 0.3$ and $c = 4.8$)

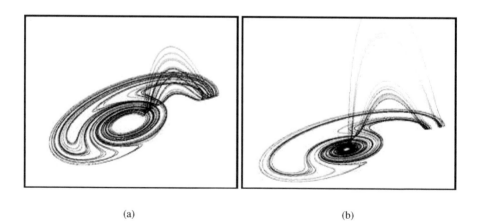

(a)   (b)

FIGURE 6.11:   Modified Rössler models

The influence of the rotation appears in Figure 6.11(a). There is a complicated movement, first in the (x, y) plane and then in the positive z direction and back, and finally the trajectories continue in the (x, y) plane. The parameters in this case were $e = 0.12$, $f = 0.4$ and $m = 5.7$. A fourth order Runge-Kutta procedure was used, with an integration step $d = 0.005$.

A different set of parameters for the model, $e = 0.2$, $f = 0.2$ and $m = 4.6$, gives the paths presented in Figure 6.11(b). The movement in the z direction now tends to higher limits, whereas the paths around the origin follow concentric elliptical orbits.

A better three-dimensional illustration appears in Figure 6.12, where we increased f to 0.3.

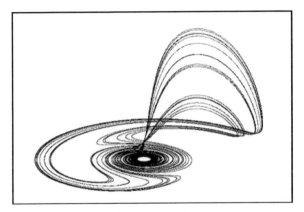

FIGURE 6.12: Another view of the modified Rössler model

6.7 The Lorenz Model

In 1963, Edward Lorenz (Lorenz, 1963) proposed an idealized model of a fluid like air or water; the warm fluid below rises, while cool fluid above sinks, setting up a clockwise or counterclockwise current. The *Prandtl number* σ, the *Rayleigh number* r and the *Reynolds number* b are parameters of the system. The width of the flow rolls is proportional to b. x is proportional to the circulatory fluid flow velocity. If $x > 0$ the fluid rotates clockwise. y is proportional to the temperature difference between ascending and descending fluid elements, and z is proportional to the distortion of the vertical temperature profile from its equilibrium. The equations of this three-dimensional model are:

$$\dot{x} = -\sigma x + \sigma y$$
$$\dot{y} = -xz + rx - y \tag{6.14}$$
$$\dot{z} = \quad xy - bz$$

The origin, $(0,0,0)$, which corresponds to a fluid at rest, is an equilibrium point for all r. For $0 < r < 1$, it is stable, while, for $r \geq 1$, it is unstable. Two other equilibrium points exist when $r \geq 1$:

$$c_+ = \left(\quad \sqrt{b(r-1)}, \quad \sqrt{b(r-1)}, r-1 \right)$$
$$c_- = \left(-\sqrt{b(r-1)}, -\sqrt{b(r-1)}, r-1 \right) \tag{6.15}$$

These represent convective circulation (clockwise and counterclockwise flow). At the equilibrium points $c_\pm$, the Lorenz model has two purely imaginary eigenvalues:

$$\lambda = \pm i \sqrt{\frac{2\sigma(\sigma+1)}{\sigma-b-1}}$$

when

$$r = r_h = \frac{\sigma(\sigma+b+3)}{\sigma-b-1}.$$

Using the standard settings

$$\sigma = 10$$
$$b = \frac{8}{3} \tag{6.16}$$

the *Hopf bifurcation point*[1] is

$$r_h = 24\frac{14}{19} \approx 24.73684$$

For $r_h < r$, all three equilibrium points of the Lorenz model are unstable. Lorenz found, numerically, that the system behaves "chaotically" whenever the Rayleigh number r exceeds the critical value r_h: All solutions are sensitive to the initial conditions, and almost all of them are neither periodic solutions nor convergent to periodic solutions or equilibrium points.

A three-dimensional view of the Lorenz attractor appears in Figure 6.13(a). Two small circles correspond to the unstable equilibrium points $c_\pm$. The parameters take the standard values (6.16). The vertical axis is z and the horizontal axis is y.

In Figure 6.13(b), a two-dimensional (x, z) view of the Lorenz attractor is presented. This is the most common illustration of this famous attractor. Perhaps the famous "butterfly effect" associated with the presence of chaos owes its name to this figure.

---

[1] See Strogatz (1994); Kuznetsov (2004).

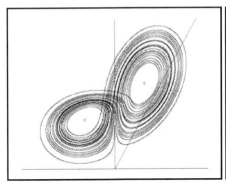

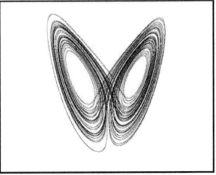

(a) A three-dimensional view   (b) The butterfly attractor

FIGURE 6.13: The Lorenz attractor

6.7.1 The modified Lorenz model

A modified Lorenz model is given by the following system of equations:

$$\dot{x} = -\sigma x + \sigma y$$
$$\dot{y} = rx - xz \qquad (6.17)$$
$$\dot{z} = xy - bz$$

The only way (6.17) differs from the original Lorenz model is that the term $-y$ is excluded from the second equation. Thus, the resulting model is simpler.

The equilibrium points are the solutions of the following system of equations:

$$-\sigma x + \sigma y = 0$$
$$rx - xz = 0 \qquad (6.18)$$
$$xy - bz = 0$$

The characteristic matrix of the modified Lorenz model is

$$\mathbf{A} - \lambda \mathbf{I} = \begin{bmatrix} -\sigma - \lambda & \sigma & 0 \\ r - z & -\lambda & -x \\ y & x & -b - \lambda \end{bmatrix}$$

Two equilibrium points appear, located at:

$$M^+ = \left(\sqrt{br}, \quad \sqrt{br}, r \right)$$
$$M^- = \left(-\sqrt{br}, -\sqrt{br}, r \right) \qquad (6.19)$$

The third equilibrium point, $M = (x, y, z) = (0, 0, 0)$, is easily seen to be unstable.

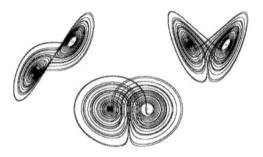

FIGURE 6.14:   The modified Lorenz model — left to right: (x, y), (y, z) and
(x, z) views

The chaotic paths of the modified Lorenz model, for $s = 10$, $b = \frac{8}{3}$ and $r = 27$,
are shown in Figure 6.14. Three two-dimensional chaotic graphs appear; from left to
right, they are the (x, y), (y, z) and (x, z) views of the attractor.
    The image on the left of Figure 6.15(a) illustrates the (x, y) view of the modified
Lorenz attractor. The image on the right presents the Poincaré (x, y) diagram of the
same attractor resulting when the plane $Z = r$ cuts the chaotic paths. This graph
gives an illustrative example of the bifurcation.

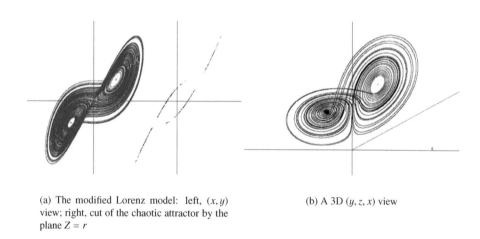

(a) The modified Lorenz model: left, (x, y)
view; right, cut of the chaotic attractor by the
plane $Z = r$

(b) A 3D (y, z, x) view

FIGURE 6.15:   The modified Lorenz chaotic attractor

Figure 6.15(b) illustrates a three-dimensional (y, z, x) view of the modified Lorenz
chaotic attractor. The two stationary points are visible. There appear to be many
similarities with the Lorenz chaotic paths presented earlier in this chapter.

Questions and Exercises

1. Show that the linearisation scheme described in section 6.1 is not applicable to the system:

$$\dot{x} = y$$
$$\dot{y} = -x^3$$

Show, furthermore, that the system is stable at the origin.

2. The system

$$\dot{x} = -y$$
$$\dot{y} = x$$

is stable at the origin, while the system

$$\dot{x} = y$$
$$\dot{y} = x$$

is not. Explain why this is so.

3. Linearise the system

$$\dot{x} = \mu x - y - x\left(x^2 + y^2\right)$$
$$\dot{y} = x + \mu y - y\left(x^2 + y^2\right)$$

at $(0, 0)$, and determine the stability of the system at the origin.

4. Consider the system:

$$\dot{x} = y$$
$$\dot{y} = -x^2$$

(a) Characterise the stability of the system at the origin.

(b) Solve the system and draw various level curves.

5. Consider the system:

$$\dot{x} = -2xy$$
$$\dot{y} = 2xy - y$$

(a) Solve the system, and check the stability at the fixed point.

(b) Draw the level curves, and discuss the form of these curves for various values of the integration parameter.

6. (a) Find the fixed points and stability of the system:

$$\dot{x} = y$$
$$\dot{y} = -\sin(x)$$

(b) Compare the behaviour of the system to that of the system:

$$\dot{x} = y$$
$$\dot{y} = -\sin(x)$$

(c) In both cases, replace $\sin(x)$ with its first-order Taylor approximation, and examine the resulting systems.

(d) What is the differences when we use the third-order Taylor approximation for $\sin(x)$?

7. Let $b > 0$. Find the fixed points of the system

$$\dot{x} = y$$
$$\dot{y} = -bx(x-1)$$

and determine the stability of the system at the points.

8. Compute the eigenvalues of the Lorenz model.

9. Consider the system:

$$\dot{x} = a - (b+1)x + x^2y$$
$$\dot{y} = \qquad bx - x^2y$$

(a) Find the equilibrium points of the system.

(b) Determine the system's stability at these points.

(c) Analyse the special cases when $b = 1$ and $b = -1$.

10. Show that when $c - ab > 0$, equation (6.12) has at least one positive root.

11. Determine if the equilibrium points M^+, M^- for the Lorenz model, given by equation (6.19), are stable.

12. Explain the butterfly effect.

Chapter 7

Non-Chaotic Systems

In this chapter we discuss various aspects of non-chaotic systems, that will be helpful in analysing other, more complex, systems. In section 7.1 we define conservative and Hamiltonian systems, while in section 7.2 we give a brief account of linear systems of differential equations, and the qualitative behaviour of their solutions near the equilibrium point. These are models for the local behaviour of more complicated systems near their equilibrium points.

In the remaining sections, we look at a number of interesting conservative systems.

7.1 Conservative Systems

In this section we consider two-dimensional *conservative systems*, that is, systems that have a non-trivial first integral of motion. For a system

$$\dot{x} = F(x, y, t)$$
$$\dot{y} = G(x, y, t) \tag{7.1}$$

A *first integral of motion* is a function $f(x, y)$, of x and y, that is independent of time. As a consequence:

$$\frac{\mathrm{d}}{\mathrm{d}t} f(x, y) = \dot{x} \frac{\partial f}{\partial x} + \dot{y} \frac{\partial f}{\partial y} = 0 \tag{7.2}$$

There is a direct connection between the level curves of the first integral and the trajectories of the system: If f is a first integral of the system (7.1), then f is constant on every trajectory of the system. Thus, every trajectory is part of some level curve of f, determined by the value of f at the initial conditions (x_0, y_0). Hence, each level curve of f is a union of trajectories.

The first integral derives its name from the usual method of computing it, by direct integration of the differential equation:

$$\frac{\mathrm{d}y}{\mathrm{d}x} = \frac{G}{F} \tag{7.3}$$

This follows from (7.2), as follows. We consider a particular level curve of f, $f(x, y) = c$, and consequently we think of f as an implicit equation for y as a function of x. In that case, we have

$$\frac{\partial f}{\partial x} + \frac{\partial f}{\partial y}\frac{dy}{dx} = 0$$

which gives us:

$$\frac{d}{dt}f(x,y) = \dot{x}\frac{\partial f}{\partial x} + \dot{y}\frac{\partial f}{\partial y} = -\dot{x}\frac{dy}{dx}\frac{\partial f}{\partial y} + \dot{y}\frac{\partial f}{\partial y} = \frac{\partial f}{\partial y}\left(-\dot{x}\frac{dy}{dx} + \dot{y}\right) \quad (7.4)$$

Equating the last term to zero, and using equation (7.1), we end up with equation (7.3). The value of the integration parameter is directly related to the level of the curve.

Since F, G in general depend on t, this integral may not be independent of t. However, in the case where F, G are independent of the time t, integration of (7.3) will result in a first integral of motion for the system. We will examine a number of particular such systems in this chapter.

Before we proceed, let us elaborate a bit more on the equations. As we saw in (7.2), the first integral of motion, $f(x, y)$, and consequently the implicit equations for the trajectories of the system, are determined by the equation:

$$F\frac{\partial f}{\partial x} + G\frac{\partial f}{\partial y} = 0 \quad (7.5)$$

Note that this equation will remain true if F, G are both multiplied by the same function of x, y. To see this geometrically, notice that the trajectories are determined completely by the direction of their tangent vector at any point, and this direction is that of the vector $(\dot{x}, \dot{y}) = (F, G)$. Therefore, multiplying both F, G with the same factor has no effect on this direction.

This means that there is a family of conservative systems sharing the same first integral of motion. From all these systems, there is one that stands out. Note that (7.5) can be rewritten as:

$$\frac{F}{\frac{\partial f}{\partial y}} = -\frac{G}{\frac{\partial f}{\partial x}}$$

Rescaling F, G amounts to selecting a value for this ratio, so a natural choice would be to set it to 1, in which case F, G, and hence the solutions of the system, are completely determined from f by:

$$F = \frac{\partial f}{\partial y} \text{ and } G = -\frac{\partial f}{\partial x} \quad (7.6)$$

In this case, we will call f the *Hamiltonian* of the system, and we will call the system a *Hamiltonian system*. All the systems we will consider in this section are Hamiltonian.

Equations (7.6) provide a necessary condition for the Hamiltonian to exist, since they imply that:

$$\frac{\partial F}{\partial x} = -\frac{\partial G}{\partial y} = \frac{\partial^2 H}{\partial x \partial y}$$

In the case where F, G are independent of t, this is also sufficient, and the Hamiltonian can be computed by simply solving the system (7.6).

Hamiltonian systems play a central role in Hamiltonian Mechanics. The equations of motion of Hamiltonian systems can be completely described using their Hamiltonian, from equations (7.5) and (7.6).

7.1.1 The simplest conservative system

The simplest conservative system is given by

$$\dot{x} = y \\ \dot{y} = -x \tag{7.7}$$

The corresponding differential equation for the integral of motion is

$$\frac{dy}{dx} = -\frac{x}{y} \tag{7.8}$$

The solutions to (7.8) are concentric circles

$$x^2 + y^2 = h$$

where h is a positive constant (Figure 7.1). Consequently, the system is conservative. Its first integral of motion is:

$$H(x, y) = \frac{1}{2} \left(x^2 + y^2 \right)$$

The level curves are therefore concentric circles, whose radius is determined by the initial conditions (x_0, y_0).

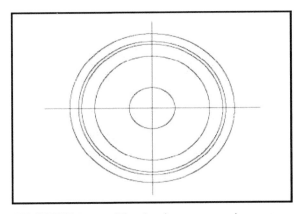

FIGURE 7.1: The simplest conservative system

Note that this is indeed a Hamiltonian system, as:

$$\frac{\partial H}{\partial y} = y = \dot{x} \text{ and } -\frac{\partial H}{\partial x} = -x = \dot{y}$$

We could have come to the same conclusions by solving the system:

$$\frac{\partial H}{\partial x} = \dot{x} = y$$

$$\frac{\partial H}{\partial y} = -\dot{y} = x$$

Integrating the first equation would yield $H(x, y) = \frac{1}{2}y^2 + c(x)$, and substituting into the second equation would yield $c(x) = \frac{1}{2}x^2 + c$.

7.1.2 Equilibrium points in Hamiltonian systems

Before we proceed to more complicated examples, it is worthwhile discussing the behaviour of a system in a neighbourhood of an equilibrium point. *Equilibrium points* are found by setting $\dot{x}$ and $\dot{y}$ to 0. In the case of a Hamiltonian system, this corresponds to

$$\frac{\partial H}{\partial x} = \frac{\partial H}{\partial y} = 0$$

So the equilibrium points of a Hamiltonian system are precisely the critical points of the Hamiltonian H. For instance, in the system 7.7 the origin is the only equilibrium point.

If (x_0, y_0) is an equilibrium point, then the Taylor expansion of H around (x_0, y_0) will be:

$$H(x, y) = H(x_0, y_0) + \frac{1}{2}h_{xx}(x - x_0)^2 + h_{xy}(x - x_0)(y - y_0) + \frac{1}{2}h_{yy}(y - y_0)^2 + \cdots \quad (7.9)$$

where

$$h_{xx} = \frac{\partial^2 H}{\partial x^2}(x_0, y_0), \qquad h_{xy} = \frac{\partial^2 H}{\partial x \partial y}(x_0, y_0), \qquad h_{yy} = \frac{\partial^2 H}{\partial y^2}(x_0, y_0)$$

are the second-order partial derivatives of H at (x_0, y_0), and the terms ignored are of higher order in $x - x_0, y - y_0$. If, for simplicity, we change coordinates so that $(x_0, y_0) = (0, 0)$, and we further denote $h_0 = H(x_0, y_0) = H(0, 0)$, then the equation (7.9) becomes:

$$H(x, y) = h_0 + \frac{1}{2}h_{xx}x^2 + h_{xy}xy + \frac{1}{2}h_{yy}y^2 + \cdots$$

So, the behaviour of the trajectories of the system, *i.e.* of the level curves of H, near the equilibrium point, is determined by the quadratic form:[1]

$$\frac{1}{2}\left(h_{xx}x^2 + 2h_{xy}xy + h_{yy}y^2\right) \tag{7.10}$$

Depending on the number of real roots of the characteristic polynomial[2] $k(X) = h_{xx} + 2h_{xy}X + h_{yy}X^2$, these trajectories may look elliptic, hyperbolic, or parabolic.[3]

If $k(X)$ has no roots, then the trajectories near the equilibrium point are elliptical, and the point is a stable equilibrium point. If $k(X)$ has two distinct roots, then the trajectories are hyperbolas, and their asymptotes are the lines with slopes the two roots (this will be elaborated further in the next section).

For instance, the system (7.7) has an equilibrium point at $(0,0)$, whereupon H is written, locally, as:

$$\frac{1}{2}x^2 + \frac{1}{2}y^2$$

In this case, the polynomial in question is $1 + X^2$, which has no real roots. Hence, the system will exhibit elliptical orbits near $(0,0)$.[4]

7.2   Linear Systems

Near an equilibrium point (we will assume this point is $0,0$), system (7.1) can be approximated by a linear system:

$$\begin{aligned}\dot{x} &= ax + by \\ \dot{y} &= cx + dy\end{aligned} \tag{7.11}$$

where a, b, c and d are real numbers or, in a more general analysis, real functions of t of a special form.

We will now discuss in greater detail the classical theory of system (7.11). This system has, as already mentioned, an equilibrium point at $(x, y) = (0,0)$. We will

---

[1] This quadratic form is the quadratic form associated to the Hessian matrix:

$$\begin{pmatrix} \frac{\partial^2 H}{\partial x^2} & \frac{\partial^2 H}{\partial x \partial y} \\ \frac{\partial^2 H}{\partial y \partial x} & \frac{\partial^2 H}{\partial y^2} \end{pmatrix}$$

[2] This number depends of course on the discriminant of the polynomial, which is the determinant of the Hessian matrix.

[3] That is, in the case that the critical point is non-degenerate, *i.e.* in the case where the characteristic polynomial is not identically zero.

[4] In this case, of course, the orbits will remain elliptical even away from $0,0$.

denote by **A** the matrix of coefficients:

$$\mathbf{A} = \begin{bmatrix} a & b \\ c & d \end{bmatrix}$$

The eigenvalues of this matrix are called the *eigenvalues* of system (7.11), and they are given by the equation

$$|\mathbf{A} - \lambda \mathbf{I}| = \begin{vmatrix} a - \lambda & b \\ c & d - \lambda \end{vmatrix} = 0$$

or, in simpler terms:

$$\lambda^2 - (a + d)\lambda + (ad - bc) = 0$$

If we note, that $a + d$ and $ad - bc$ are respectively the *trace* and *determinant* of **A**, we can write the equation for the eigenvalues as:

$$\lambda^2 - \text{tr}(\mathbf{A})\lambda + \det(\mathbf{A}) = 0$$

The solutions to this are

$$\lambda_{1,2} = \frac{\text{tr}(\mathbf{A}) \pm \sqrt{\Delta}}{2} \qquad (7.12)$$

where, as usual, Δ denotes the discriminant:

$$\Delta = \text{tr}(\mathbf{A})^2 - 4\det(\mathbf{A}) \qquad (7.13)$$

Before we proceed to a classification of equilibrium points according to the qualitative behaviour of the solutions near these points, we will discuss the effects of various transformations on the shape of the matrix A.

7.2.1   Transformations on linear systems

We start by considering a simple *coordinate rescaling*:

$$u = kx \, v \quad = k'y \qquad (7.14)$$

The transformed system (7.11) then takes the form

$$\dot{u} = \quad au + rbv$$

$$\dot{v} = \frac{1}{r}cu + \quad dv$$

where $r = \frac{k}{k'}$. So the matrix of the new system, **A'**, is:

$$\mathbf{A'} = \begin{bmatrix} a & rb \\ \frac{1}{r}c & d \end{bmatrix}$$

The trace and determinant thus remain invariant:

$$\text{tr}(\mathbf{A}') = \text{tr}(\mathbf{A})$$
$$\det(\mathbf{A}') = \det(\mathbf{A})$$

A more general *change of coordinates* amounts to the transformation

$$\begin{bmatrix} u & v \end{bmatrix} = \mathbf{C} \begin{bmatrix} x & y \end{bmatrix} \tag{7.15}$$

where $\mathbf{C}$ is an invertible matrix of constants. The corresponding matrix $\mathbf{A}'$ is then a matrix similar to $\mathbf{A}$:

$$\mathbf{A}' = \mathbf{CAC}'$$

Hence, this again keeps the trace and determinant invariant.

Another simple transformation involves *time rescaling*:[5]

$$u(t) = x(kt) \quad v(t) = y(kt) \tag{7.16}$$

System (7.11) then becomes:

$$\dot{u} = kau + kbv$$
$$\dot{v} = kcu + kdv$$

Thus the matrix of the system becomes

$$\mathbf{A}' = k\mathbf{A}$$

and hence

$$\text{tr}(\mathbf{A}') = k\,\text{tr}(\mathbf{A})$$
$$\det(\mathbf{A}') = k^2\det(\mathbf{A})$$

According to (7.12) and (7.13), the eigenvalues are simply rescaled by k.

Finally, the transformation

$$u(t) = e^{-kt}x(t) \quad v(t) = e^{-kt}y(t) \tag{7.17}$$

results in a system with matrix

$$\mathbf{A}' = \mathbf{A} - k\mathbf{I}$$

with the corresponding changes this implies to the trace and determinant:

$$\text{tr}(\mathbf{A}') = \text{tr}(\mathbf{A}) - 2k$$
$$\det(\mathbf{A}') = k^2 - k\,\text{tr}(\mathbf{A}) + \det(\mathbf{A})$$

Note however, that this transformation does keep the discriminant

$$\Delta = \text{tr}(\mathbf{A})^2 - 4\det(\mathbf{A})$$

invariant, and hence it amounts to a parallel translation of the eigenvalues by $-k$. This transformation changes radically the behaviour of solutions near $(0,0)$, but in a very controlled way, and we will utilise it in the next section.

---

[5]This does not alter in any essential way the behaviour of solutions near the point $(0,0)$, provided $k > 0$.

7.2.2   Qualitative behaviour at equilibrium points

We will now proceed to analyse the qualitative behaviour of solutions near equilibrium points, using the transformations from section 7.2.1. First off, the sign of tr $\mathbf{A}$ is of paramount importance. Using transformation (7.17), we can always reduce to the case where tr($\mathbf{A}$) = 0, *i.e.* where $d = -a$, and we will do so for the next couple of paragraphs. The condition tr($\mathbf{A}$) = 0 is exactly the condition for system (7.11) to be conservative. In that case, the system has the first integral:

$$f(x, y) = cx^2 - 2axy - by^2$$

We have to distinguish a couple of cases here.

1. $\Delta = 0$, *i.e.* the first integral f is degenerate, and there is only one, double, eigenvalue, namely 0. Up to a change of coordinates, we may assume $f(x, y) = x^2$, *i.e.* that the linear system is:

$$\dot{x} = 0$$
$$\dot{y} = x$$

   The solutions are then:

$$x(t) = A$$
$$y(t) = At + B$$

   This case is illustrated in Figures 7.2(a) and 7.2(b).

2. $\Delta < 0$. Then, there are two complex eigenvalues, which, after performing transformation (7.16), we may assume to be $\pm i$. Then, up to a change of coordinates, we may assume $f(x, y) = x^2 + y^2$, *i.e.* that the linear system is

$$\dot{x} = -y$$
$$\dot{y} = \quad x$$

   The solutions are then:

$$x(t) = A\cos(t) + B\sin(t)$$
$$y(t) = B\cos(t) - A\sin(t)$$

   This case is illustrated in Figures 7.2(c) and 7.2(d).

3. $\Delta > 0$. Then there are two distinct real eigenvalues, which, after performing tranformation (7.16), we may assume to be ± 1. After a change of coordinates, we may assume that $f(x, y) = xy$, *i.e.* that the system is:

$$\dot{x} = x$$
$$\dot{y} = -y$$

   The solutions are then:

$$x(t) = Ae^t$$
$$y(t) = Be^{-t}$$

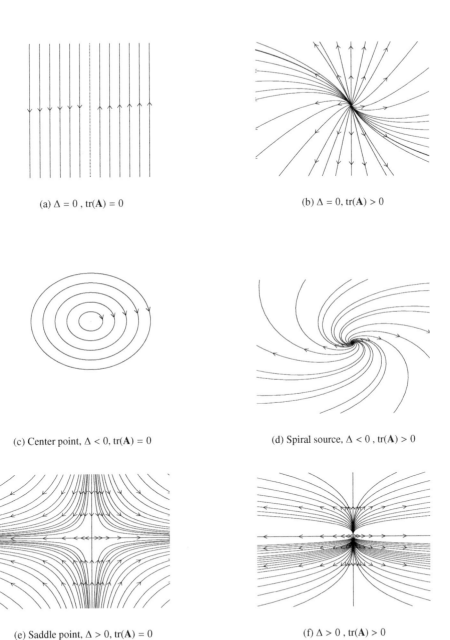

(a) $\Delta = 0$, tr(**A**) = 0

(b) $\Delta = 0$, tr(**A**) > 0

(c) Center point, $\Delta < 0$, tr(**A**) = 0

(d) Spiral source, $\Delta < 0$, tr(**A**) > 0

(e) Saddle point, $\Delta > 0$, tr(**A**) = 0

(f) $\Delta > 0$, tr(**A**) > 0

FIGURE 7.2:   Qualitative classification of equilibrium points

This case is illustrated in Figures 7.2(e) and 7.2(f).

To summarise, the behaviour of solutions near an equilibrium point falls under one of the following cases:

1. $\Delta = 0$. This is the degenerate case.

2. $\Delta > 0$. Then $\lambda_{1,2}$ are both real. The behaviour in this case depends a lot on their sign. If they are both negative, then all solutions tend toward the equilibrium point, which in this case is called a *stable point*, or also a *stable node*. If they are both positive, then the solutions tend away from the equilibrium point, and the point is classified as *unstable*.

   Otherwise, $\lambda_{1,2}$ are both real, but with opposite signs. In this case, there is a stable direction and an unstable direction, indicated by the negative and positive eigenvalues respectively. The equilibrium point is then called a *saddle point*.

3. $\Delta < 0$. Then $\lambda_{1,2}$ have non-zero real and imaginary parts. If the real part is negative, then the solutions spiral toward the equilibrium point, which is called in this case a *stable spiral* or a *spiral sink*. If, on the other hand, the real part is positive, then the solutions spiral away from the critical point, and the point is then called an *unstable point* or a *spiral source*.

   A special case is when $\lambda_{1,2}$ have only an imaginary part (*i.e.* when the real part is zero). The critical point is then called a *center*.

7.3   Egg-Shaped Forms

7.3.1   A simple egg-shaped form

We now proceed to examine certain Hamiltonian systems that give rise to very interesting egg-shaped forms. The first such system is:

$$\dot{x} = y(y - b)$$
$$\dot{y} = x \tag{7.18}$$

This is a Hamiltonian system, since

$$\frac{\partial}{\partial x}\big(y(y - b)\big) = 0 = -\frac{\partial}{\partial y}(x)$$

The Hamiltonian is easily seen to be

$$H(x, y) = -\frac{x^2}{2} - b\frac{y^2}{2} + \frac{y^3}{3} + h$$

This system has two equilibrium points, at

$$(x, y) = (0, 0) \text{ and } (x, y) = (0, b)$$

The first of these is a stable point, as the quadratic form,

$$-\frac{1}{2}\left(x^2 + by^2\right)$$

has two imaginary roots:

$$\lambda_{1,2} = \pm\frac{i}{\sqrt{b}}$$

The second point, however, is unstable, since its characteristic polynomial,

$$-2 + 2bX^2 = 2(\sqrt{b}X - 1)(\sqrt{b}X + 1)$$

has two real roots:

$$\lambda = \pm\frac{1}{\sqrt{b}}$$

These are exactly the slopes of the two tangent lines at $(0, b)$.

For convenience, we will choose the integration constant in H to be such, that H is 0 at this, second, equilibrium point, namely:

$$h = \frac{b^3}{6}$$

The level curve that passes through $(0, b)$ is shown, marked with a heavy line, in Figure 7.3(a) ($b = 0.6$). It has equation:

$$H(x, y) = -\frac{x^2}{2} - b\frac{y^2}{2} + \frac{y^3}{3} + \frac{b^3}{6}$$

The sharp (top) corner of the egg-shaped form is located at the equilibrium point $(0, b)$, whereas the bottom of the form is at the point $\left(0, -\frac{b}{2}\right)$.[6] At this point, the speed of the particle is given by $\dot{x} = \frac{3}{4}b^2$. The boundaries of the egg-shaped form in the x direction are when $y = 0$, namely $x = \pm\sqrt{\frac{b^3}{3}}$. The speed at these points is $\dot{y} = \pm\sqrt{\frac{b^3}{3}}$ respectively.

7.3.2 A double egg-shaped form

A double egg-shaped form with interconnected sharp corners is expressed by the following system:

$$\begin{aligned} \dot{x} &= -y(y - a)(y - b) \\ \dot{y} &= x \end{aligned} \tag{7.19}$$

---

[6]This is found by solving the cubic polynomial $H(0, y) = 0$ This will have a double root at $y = b$, and one more root.

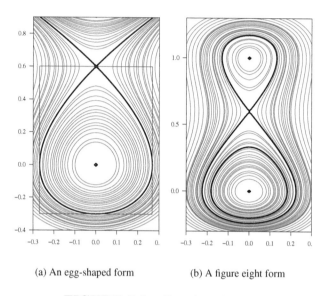

(a) An egg-shaped form   (b) A figure eight form

FIGURE 7.3:   Egg-shaped forms

where we will assume that $0 < b < a$. The Hamiltonian in this case is[7]

$$H(x,y) = -\frac{y^4}{4} + (a+b)\frac{y^3}{3} - ab\frac{y^2}{2} - \frac{x^2}{2} + \frac{b^3}{12}(b-2a)$$

Figure 7.3(b) shows the trajectories when $a = 1$ and $b = 0.6$.

The first egg-shaped form at the top of the figure has a stable equilibrium point at $(a, 0)$, whereas the other egg-shaped form has a stable equilibrium point located at $(0, 0)$. An unstable point is located at $(0, b)$, where the sharp corners of the two egg-shaped forms coincide.

When the initial values are $x_0 = 0$ and $y_0 = -b$, a figure eight shape appears. This is a characteristic level curve marked by a heavy line in Figure 7.3(b). The integration constant has the value $h = -\frac{b^4}{12} + \frac{ab^3}{6}$ for this particular case. Inside this form the trajectories follow closed loops, and the same holds outside this characteristic level.

7.3.3   A double egg-shaped form with an envelope

A double egg-shaped form enclosed in an external envelope is provided by the following system:

$$\dot{x} = y(y^2 - b^2)(y + a)$$
$$\dot{y} = x \tag{7.20}$$

---

[7]The constant was again chosen, so that the Hamiltonian is zero at the unstable point.

The Hamiltonian of this system is given by:

$$H(x, y) = \frac{y^5}{5} + a\frac{y^4}{4} - b^2\frac{y^3}{3} - ab^2\frac{y^2}{2} - \frac{x^2}{2}$$

where we've set the integration constant to zero for simplicity. Figure 7.4 illustrates this case where $a = 1$ and $b = 0.6$. The first egg-shaped form at the top of the figure has a stable equilibrium point at $(0, 0)$, whereas the second egg-shaped form has a stable equilibrium point at $(0, -a)$. There are two unstable points, one at $(0, -b)$, where the sharp corners of the two egg-shaped forms coincide, and one at $(0, b)$, where the envelope is formed. The level curves through those two unstable points are marked with a heavy line in the graph. For this particular case, corresponding values of the Hamiltonian on these curves are

$$\begin{aligned}
h_{b-} &= -\frac{2}{15}b^5 - \frac{1}{4}ab^4 \\
h_{b+} &= \frac{2}{15}b^5 - \frac{1}{4}ab^4
\end{aligned}$$

(7.21)

Inside the envelope, the trajectories follow closed loops, whereas, outside of it, the trajectories diverge to infinity.

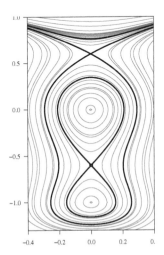

FIGURE 7.4: A double egg-shaped form with an envelope

7.4 Symmetric Forms

We consider now certain forms that exhibit symmetry. A simple symmetric Hamiltonian form is based on the system:

$$\dot{x} = y(y - b)$$
$$\dot{y} = x(x - b)$$

(7.22)

The corresponding Hamiltonian is

$$H(x, y) = \frac{y^3 - x^3}{3} - b\frac{y^2 - x^2}{2}$$

The level curves at level 0 are marked by a thick line in Figure 7.5(a). There are two stable points, at $(b, 0)$ and $(0, b)$, and two unstable points, at $(0, 0)$ and (b, b).

A slight change to (7.22) gives rise to another double-shaped form, with symmetry with respect to the $y = x$ axis. The equations are:

$$\dot{x} = -y(y - b)$$
$$\dot{y} = x(x - b)$$

(7.23)

The Hamiltonian is

$$H(x, y) = \frac{y^3 + x^3}{3} - b\frac{y^2 + x^2}{2}$$

The level curve for the outer periphery has level $-\frac{b^3}{6}$, and is illustrated in Figure 7.5(b). The stable equilibrium points are located at $(0, 0)$ and (b, b), whereas the two unstable points are located at $(0, b)$ and $(b, 0)$. The three points of the level curve located on the line $x = y$ are: the point $\left(\frac{b}{2}, \frac{b}{2}\right)$, and the other two solutions of the equation $4x^3 - 6bx^2 + b^3 = 0$.

A simple non-symmetric form is expressed by the following two equations:

$$\dot{x} = -y(y - b)$$
$$\dot{y} = x(x - a)$$

(7.24)

The Hamiltonian in this case is

$$H(x, y) = \frac{y^3 + x^3}{3} - \frac{by^2 + ax^2}{2}$$

There are two characteristic level curves, passing through the unstable points $(a, 0)$ and $(0, b)$ respectively, and with corresponding levels $h_a = -\frac{1}{6}a^3$ and $h_b = -\frac{1}{6}b^3$ respectively. These level curves are shown in Figure 7.6, marked with a thick line. There are two stable equilibrium points, at $(0, 0)$ and (a, b).

An interesting three-fold shape is given by the system

$$\dot{x} = y(y^2 - b^2)$$
$$\dot{y} = x(x - b)$$

(7.25)

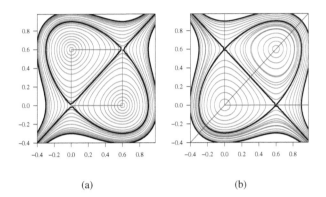

(a)   (b)

FIGURE 7.5:   Symmetric forms

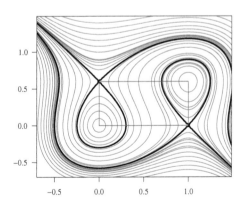

FIGURE 7.6:   A simple non-symmetric form

The corresponding level curves of the Hamiltonian

$$\frac{y^4}{4} - b^2\frac{y^2}{2} - \frac{x^3}{3} + b\frac{x^2}{2}$$

are shown in Figure 7.7(a) ($b = 0.85$).

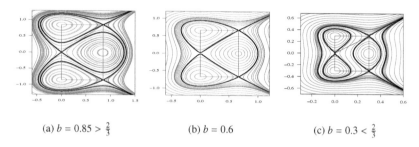

(a) $b = 0.85 > \frac{2}{3}$　　　　　(b) $b = 0.6$　　　　　(c) $b = 0.3 < \frac{2}{3}$

FIGURE 7.7:   A symmetric three-fold form

There are two characteristic levels for H, which provide the level curves marked by a heavy line. The first value, $h = 0$, corresponds to the equilibrium point $(0,0)$, and gives the central formation. This curve also passes through the point $(\frac{3}{2}b, 0)$ where x takes its maximum value in the bounded part of this curve.

The second characteristic level is $h = \frac{b^3}{6} - \frac{b^4}{4}$, for which the level curve passes through the two unstable equilibrium points $(b, \pm b)$. There are three stable equilibrium points, two at $(0, \pm b)$, where $H = -\frac{b^4}{4}$, and one at $(b, 0)$, where $H = \frac{b^3}{6}$.

The shape varies a lot depending on the parameter b. When $b > \frac{2}{3}$, we get a figure similar to Figure 7.7(a). When $b = \frac{2}{3}$, the two characteristic curves coincide, and the corresponding graph is shown in Figure 7.7(b). When $b < \frac{2}{3}$, the two curves are not so interrelated (Figure 7.7(c)).

7.5   More Complex Forms

More complex forms appear by adding new terms in the right-hand side of the Hamiltonian equations of the two-dimensional system. The four-type symmetric form illustrated in Figure 7.8(a) corresponds to the system

$$\begin{aligned}
\dot{x} &= -y(y^2 - b^2) \\
\dot{y} &= x(x^2 - b^2)
\end{aligned}$$

(7.26)

with Hamiltonian:

$$\frac{x^4 + y^4}{4} - b^2 \frac{x^2 + y^2}{2}$$

There are five stable equilibrium points, one at the origin $(0,0)$, and the other four located at the symmetric points (b,b), $(b,-b)$, $(-b,b)$ and $(-b,-b)$. There are also four unstable equilibrium points located at $(0,b)$, $(0,-b)$, $(b,0)$ and $(-b,0)$. The level curve passing through those equilibrium points has level $-\frac{b^4}{4}$. In Figure 7.8(a), a heavy line marks this critical level curve.

Introducing a new parameter a into (7.26), the following system arises:

$$\begin{aligned}\dot{x} &= -y(y^2 - b^2)\\ \dot{y} &= x(x^2 - a^2)\end{aligned} \qquad (7.27)$$

Its Hamiltonian is

$$\frac{x^4 + y^4}{4} - \frac{a^2 x^2 + b^2 y^2}{2}$$

This system has a richer structure, as illustrated in Figure 7.8(b) $(a = 0.7, b = 0.5)$. Two distinct level curves appear, for values $H = -\frac{1}{4}a^4$ and $H = -\frac{1}{4}b^4$ respectively. The first curve is associated with two 8-shape forms of Figure 7.8(b). The second curve is associated with the other more complicated form, also marked with a heavy line. The four points located on the x axis $(y = 0)$ are given by:

$$x = \pm\sqrt{a^2 \pm \sqrt{a^4 - b^4}}$$

If the parameter a is introduced in both equations, then an interesting system arises:

$$\begin{aligned}\dot{x} &= -y(y - a)(y - b)\\ \dot{y} &= x(x - a)(x - b)\end{aligned} \qquad (7.28)$$

Its Hamiltonian is:

$$H(x,y) = \frac{x^4 + y^4}{4} - (a + b)\frac{x^3 + y^3}{3} + ab\frac{x^2 + y^2}{2}$$

The model, with parameters equal to $a = 0.8$ and $b = 0.3$, is illustrated in Figure 7.8(c).

There is an outer characteristic level curve, with level $h_b = -\frac{1}{12}b^4 + \frac{1}{6}ab^3$, and an inner characteristic level curve, with level $h_a = -\frac{1}{12}a^4 + \frac{1}{6}a^3b$. There are five stable points located at $(0,0)$, (a,a), (b,b), $(a,0)$ and $(0,a)$, and four unstable points at $(b,0)$, $(0,b)$, (a,b) and (b,a).

When $b = \frac{1}{2}a$, then $h_b = 0$, and the system leads to a perfect symmetry, illustrated in Figure 7.8(d) when the parameters are $a = 0.8$ and $b = \frac{a}{2} = 0.4$.

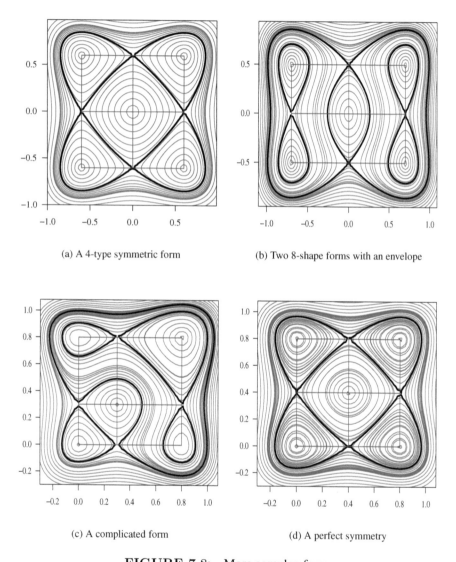

(a) A 4-type symmetric form

(b) Two 8-shape forms with an envelope

(c) A complicated form

(d) A perfect symmetry

FIGURE 7.8:   More complex forms

7.6 Higher-Order Forms

We conclude this section with a couple of much more complicated examples, arising from systems whose equations involve higher-order polynomials. One such example, illustrated in Figure 7.9(a), is the system:

$$\dot{x} = -(y - a)(y - b)(y - c)(y - e)(y - f)$$
$$\dot{y} = (x - a)(x - b)(x - c)(x - e)(x - f)$$

(7.29)

The Hamiltonian can be easily computed from the above equations, but we will omit
the tedious task here.

The parameters were set to $a = 0.1$, $b = 0.6$, $c = 0.9$, $e = 1.5$ and $f = 2$. There
are 13 stable equilibrium points, and 12 unstable equilibrium points. Equilibrium
regions and regions with high movement (flows) appear. As in the simpler cases,

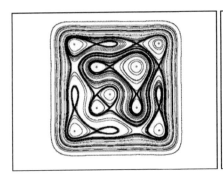 

(a) Symmetry only along $y = x$ axis   (b) Multiple symmetries

FIGURE 7.9: Higher-order forms

appropriate values for the parameters lead to symmetric forms, as in Figure 7.9(b),
where the parameters are $a = 0$, $b = 0.5$, $c = 1$, $e = 1.5$ and $f = 2$.

Questions and Exercises

1. Consider the system

$$\dot{x} = -y$$
$$\dot{y} = \ x(1 - x)$$

(a) Find its equilibrium points.

(b) Find the eigenvalues and eigenvectors at the equilibrium points, and characterise the points according to their stability.

(c) Linearise the system, and find the type of stability at the equilibrium points.

(d) Find a first integral of motion for the system.

(e) Draw various trajectories of the system in the (x, y) plane.

(f) Draw the trajectories that pass through equilibrium points.

(g) Find the maximum speed of the flow at the y direction $(\dot{y})$.

2. Consider the system

$$\dot{x} = 1 - \ y$$
$$\dot{y} = x + ay$$

(a) Find the equilibrium points for the system.

(b) Find the characteristic equation at the equilibrium points.

(c) Characterise the equilibrium points for various values of a.

(d) Examine in particular the case $a = 2$.

3. For each of the following systems, (a) show that it is Hamiltonian, (b) compute the Hamiltonian function, (c) determine the fixed points, and classify them, and (d) plot, for various values of the parameters, the level curves of the Hamiltonian, including the curves passing through the fixed points.

(a)
$$\dot{x} = y(y - b)$$
$$\dot{y} = ax$$

(b)
$$\dot{x} = y^2 - b^2$$
$$\dot{y} = x$$

(c)
$$\dot{x} = y^2 - b^2$$
$$\dot{y} = x^2 - a^2$$

4. Show that the system

$$\dot{x} = 2xy$$
$$\dot{y} = x^2 - y^2$$

is conservative, but not Hamiltonian. Find a system equivalent to it that is Hamiltonian.

5. Show that the system

$$\dot{x} = \frac{1}{x}$$

$$\dot{y} = -\frac{1}{y}$$

is conservative, but not Hamiltonian. Find a system equivalent to it that is Hamiltonian.

6. Show that each of the transformations described in section 7.2.1 have the indicated effect on the matrix $\mathbf{A}$ and its eigenvalues.

Chapter 8

Rotations

In this chapter we examine the effects of introducing affine transformations to models, in particular the effect of introducing rotation. A wealth of very interesting models arise from this seemingly simple change.

8.1 Introduction

The location (x_n, y_n) of a system is given in parametric form (r_n, θ_n) by:

$$x_n = r_n \cos(\theta_n)$$
$$y_n = r_n \sin(\theta_n) \tag{8.1}$$

where $r = \sqrt{x^2 + y^2}$ and θ_n is the rotation angle.

If we assume that transitioning from the n-th state to the $n+1$-st state amounts to a rotation by an angle $\Delta\theta$, then the new coordinates would be given by:

$$x_{n+1} = r_n \cos(\theta_n + \Delta\theta)$$
$$y_{n+1} = r_n \sin(\theta_n + \Delta\theta) \tag{8.2}$$

or after using the familiar trigonometric identities for the sum of two angles:

$$x_{n+1} = r_n(\cos(\theta_n)\sin(\Delta\theta) - \sin(\theta_n)\sin(\Delta\theta))$$
$$y_{n+1} = r_n(\cos(\theta_n)\sin(\Delta\theta) + \sin(\theta_n)\cos(\Delta\theta)) \tag{8.3}$$

Therefore, the effect of rotation can be written directly using the following, which may be familiar to you already:

$$x_{n+1} = x_n \cos(\Delta\theta) - y_n \sin(\Delta\theta)$$
$$y_{n+1} = x_n \sin(\Delta\theta) + y_n \cos(\Delta\theta) \tag{8.4}$$

When the rotation is followed by a translation equal to a and parallel to the x axis, then the transformation (8.4) is replaced by:

$$x_{n+1} = a + x_n \cos(\Delta\theta) - y_n \sin(\Delta\theta)$$
$$y_{n+1} = \quad\; x_n \sin(\Delta\theta) + y_n \cos(\Delta\theta) \tag{8.5}$$

We can consider the system (8.5) as a simple system of difference equations. Let us work out an analogous system of differential equations. The time elapsed between two successive iterations is Δt and is usually set to 1. Then, taking into account that

$$\dot{x} = \frac{dx}{dt} \approx \frac{\Delta x}{\Delta t} = x_n - x_{n-1}$$

and similarly for $\dot{y}$, we obtain the differential equations:

$$\begin{aligned} \dot{x} &= a + x(\cos(\Delta\theta) - 1) - y\sin(\Delta\theta) \\ \dot{y} &= \quad x\sin(\Delta\theta) \quad\quad + y(\cos(\Delta\theta) - 1) \end{aligned} \tag{8.6}$$

8.2   A Simple Rotation-Translation System of Differential Equations

If the rotation angle is small, $\Delta\theta \ll 1$, the system (8.6), using first order Taylor approximations for cos and *sin*, can be simplified to:

$$\begin{aligned} \dot{x} &= a - y\Delta\theta \\ \dot{y} &= \quad x\Delta\theta \end{aligned} \tag{8.7}$$

This is a Hamiltonian system, whose first integral of motion can be computed from the differential equation:

$$\frac{\partial y}{\partial x} = \frac{x\Delta\theta}{a - y\Delta\theta}$$

This equation leads to the form:[1]

$$a\,dy = (\Delta\theta)r\,dr \tag{8.8}$$

The solution of equation (8.8) depends on the form of the function $\Delta\theta$. A reasonable assumption would be that $\Delta\theta$ is "radial," *i.e.* depends only on the distance r from the origin, and not of the direction. A standard choice in mechanics, based on the transverse component of the acceleration, is that the rotational change is inversely proportional to the square distance, *i.e.* that:

$$\Delta\theta \approx \frac{h_1}{r^2}$$

Another important case is that of a mass M rotating in a circular orbit under a central force

$$f(r) = \frac{MG}{r^2}$$

---

[1] We have used here that $r^2 = x^2 + y^2$, and consequently $r\,dr = x\,dx + y\,dy$.

where G is the gravity constant. The equations of motion then result in

$$\Delta\theta \approx \dot\theta = \sqrt{\frac{MG}{r^3}}$$

In this case, equation (8.8) becomes:

$$a\,dy = \sqrt{\frac{MG}{r}}\,dr$$

Its solution is

$$r = \frac{(ay+h)^2}{4MG}$$

where h is an integration constant. This can also be rewritten as

$$x^2 = \left(\frac{(ay+h)^2}{4MG} - y\right)\left(\frac{(ay+h)^2}{4MG} + y\right) \tag{8.9}$$

The fixed point of the system is at

$$(x,y) = \left(0, \frac{MG}{a^2}\right)$$

at which point the value of h is:

$$h = \frac{MG}{a}$$

The path of the trajectory in the (x, y) plane for this value of h divides the plane in two segments. When $h > \frac{MG}{a}$, the trajectories diverge and the rotating object heads off to infinity. When $h < \frac{MG}{a}$ the rotating mass moves inside the space bounded by the above trajectory.

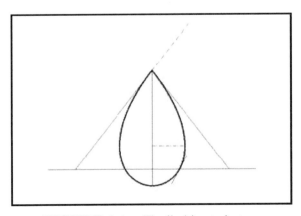

FIGURE 8.1: The limiting trajectory

Some very interesting properties of equation (8.8) are illustrated in Figure 8.1. The trajectory in the limit of escape has an egg shape, similar to but different from those discussed in Chapter 7. The sharp corner is at the maximum value of $y = \frac{MG}{a^2}$, where $x = 0$. This is the fixed point for the system, which is not stable.

The minimum value of y is that where the maximum rotation speed is achieved. This can be found by setting $x = 0$ in equation (8.9). This gives us two solutions, the fixed point at $y = \frac{MG}{a^2}$ as already discussed, and the solution to the equation:

$$(ay + h)^2 = -4ahy$$

The largest solution of this equation is:

$$y = \left(2\sqrt{2} - 3\right)\frac{MG}{a^2} \approx -0.1716\frac{MG}{a^2} \tag{8.10}$$

The maximum value of x is given by:

$$x_{max} = \frac{MG}{a^2}\sqrt{\frac{5\sqrt{5} - 11}{2}} \approx 0.3\frac{MG}{a^2}$$

This is computed by equating to zero the first derivative of x with respect to y, that is:

$$\frac{\partial x}{\partial y} = \frac{\dot{x}}{\dot{y}} = \frac{1}{x}\left(\frac{a}{8(MG)^2}\left(ay + \frac{MG}{a}\right)^3 - y\right) = 0 \tag{8.11}$$

Setting $z = \frac{a}{h}y$, equation (8.11) becomes:

$$(z + 1)^3 = 8z$$

This can be easily seen to have the three roots:

$$z = 1, \quad z = \sqrt{5} - 2, \quad z = -\sqrt{5} - 2$$

The first of these corresponds to the fixed point $\left(0, \frac{MG}{a^2}\right)$. The second is the one we are after, and the corresponding y value is:

$$y = \frac{MG}{a^2}(\sqrt{5} - 2)$$

When $y = 0$, then

$$x = \frac{MG}{4a^2} = \frac{h}{4a}, \quad r = \frac{h^2}{4MG} = \frac{h}{4a} \quad \text{and } \Delta\theta = \frac{8a^2}{h}$$

so from (8.7) we get:

$$\frac{dy}{dx} = \frac{\dot{y}}{\dot{x}} = \frac{x\Delta\theta}{a - y\Delta\theta} = \frac{\frac{h}{4a} \cdot \frac{8a^2}{h}}{a} = 2$$

This tangent appears in Figure 8.1.

The tangent at the top sharp corner of the egg-shaped path can be seen by writing equation (8.8) in a coordinate system centered at this point. For this reason, let:

$$z = y - \frac{MG}{a^2} = y - \frac{h}{a}$$

Then, in terms of (x, z), equation (8.8) can be written as:

$$0 = -x^2 + \frac{a}{4h}z^2 \cdot \left(\frac{2h}{a} + 2z + \frac{a}{4h}z^2\right) = -x^2 + \frac{1}{2}z^2 + \cdots$$

where the omitted terms are of higher order in z. This means that near $(x, z) = (0, 0)$, our orbit behaves exactly like the function:

$$0 = -x^2 + \frac{1}{2}z^2 = \left(\frac{1}{\sqrt{2}}z - x\right)\left(\frac{1}{\sqrt{2}}z + x\right)$$

So the two tangent lines are given by the equations:

$$y = \frac{h}{a} \pm \sqrt{2}x$$

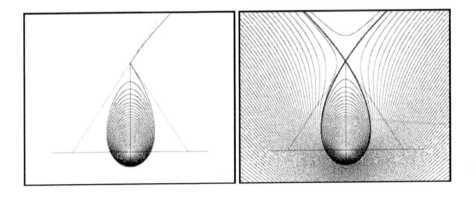

(a) Rotating paths   (b) Paths in the plane

FIGURE 8.2: Rotating paths

Figure 8.2(a) illustrates the path of a moving particle. The particle follows an anti-clockwise direction. It starts from a point near zero and rotates away from the centre until the highest permitted place. Then, it escapes to infinity following the tangent on the top of the egg-shaped path. In Figure 8.2(b), a large number of paths are drawn. Every path has the same value for $h = \frac{MG}{a^2}$. The paths outside the egg-shaped limits have direction from left to right and they do not transverse the egg-shaped region of the plane. The iterative formula used for the above simulations is:

$$x_{n+1} = x_n + \left(a - y_n \sqrt{MG} r_n^3\right) d$$
$$y_{n+1} = y_n + x_n \sqrt{MG} r_n^3 d \qquad (8.12)$$
$$r_n = \sqrt{x_n^2 + y_n^2}$$

where, for the above figures, $a = 2$, $MG = 100$ and $d = 0.05$.

Other interesting properties of the aforementioned egg-shaped forms arise by using elements from mechanics. From the equations of motion, it is clear that the parameter a is a constant speed with direction parallel to the x axis. The rotational speed has two components: $\dot{x}, \dot{y}$. However, at the top and at the bottom of the egg-shaped form only the x component of the velocity is present, since $x = 0$. The rotational speed, or the transverse component of the velocity, is $v = \sqrt{\frac{MG}{y}}$. This speed must be equal to a. Thus, without using the equations for the trajectories, the value of y at the top, $x = 0$, is estimated from the equality $v - a = 0$. Thus,

$$\sqrt{\frac{MG}{y}} = a$$

and

$$y = \frac{MG}{a^2}$$

To obtain the equation for the total speed at the lower point of the curve, we must take into account that this speed must be equal to the escape speed:

$$v + a = v_{\text{esc}} \qquad (8.13)$$

The escape speed for a mass rotating around a large body is known from mechanics to be:

$$v_{\text{esc}} = \sqrt{\frac{2MG}{y}}$$

Thus, equation (8.13) becomes:

$$\sqrt{\frac{MG}{y}} + a = \sqrt{\frac{2MG}{y}}$$

Finally, from this last equation the value of y is:

$$y = \frac{MG}{a^2}(\sqrt{2} - 1)^2$$

This is, up to a sign, the same result obtained in equation (8.10). It is obvious that the system in these particular cases obeys the laws of classical mechanics. This egg-like form is a set of vortex curves with relative stability and strength.

8.3   A Discrete Rotation-Translation Model

The analysis above is significant in helping us understand the dynamics under-lying the rotation analogue in the discrete case. The iterative scheme based on the difference equations for rotation-translation is more difficult to handle analytically, but, on the other hand, it is closer to a real life situation. No approximation is needed, whereas, the egg-shaped scheme and the trajectories are retained. Moreover, the tra-jectories inside the egg-shaped boundary are perfectly closed loops. Additionally, the chaotic nature of the rotation-translation procedure near the centre becomes appar-ent. In the equations (8.4), we introduce an area contracting parameter b $(0 < b < 1)$, and the final equations used are the following

$$\begin{aligned} x_{n+1} &= a + b\,(x_n \cos \Delta\theta - y_n \sin \Delta\theta) \\ y_{n+1} &= \quad\; b\,(x_n \sin \Delta\theta + y_n \cos \Delta\theta) \end{aligned} \qquad (8.14)$$

where the same approximation

$$\Delta\theta_n \approx \sqrt{\frac{MG}{r_n^3}}$$

is used.

A convenient formulation of the last iterative map expressing rotation and transla-tion is achieved by using the following difference equation

$$f_{n+1} = a + b f_n e^{i\Delta\theta} \qquad (8.15)$$

where $f_n = x_n + i y_n$.

The Jacobian determinant of this map is $J = b^2 = \lambda_1 \lambda_2$. This is an area contracting map if the Jacobian J is less than 1. The eigenvalues of the Jacobian are λ_1 and λ_2, and the translation parameter is a.

A very interesting property of the map (8.14) is the existence of a disk F of radius $\frac{ab}{1-b}$ centered at $(a, 0)$ in the (x, y) plane. The points inside the disk remain inside it under the map.[2]

To locate the fixed points of this transformation, let us change coordinates by setting:

$$z_n = f_n - \frac{a}{1 - b^2}$$

Then the transformation can be rewritten as:

$$z_{n+1} - z_n = \left(b e^{i\Delta\theta} - 1\right) z_n + ab \left(\frac{e^{i\Delta\theta} - b}{1 - b^2}\right)$$

---

[2]This fact is left as an exercise for the reader.

So for a fixed point we would have:

$$z = \frac{ab}{1-b^2} \cdot \left(\frac{e^{i\Delta\theta} - b}{1 - be^{i\Delta\theta}} \right) \tag{8.16}$$

It can easily be verified that

$$\left| \frac{e^{i\Delta\theta} - b}{1 - be^{i\Delta\theta}} \right| = 1$$

and consequently all fixed points lie on the circle centered at $\left(\frac{a}{1-b^2}, 0 \right)$ and with radius $\frac{ab}{1-b^2}$, in other words they satisfy:

$$\left(x - \frac{a}{1-b^2} \right)^2 + y^2 = \frac{a^2 b^2}{(1 - b^2)^2}$$

Further, if we transform equation (8.16) back to an equation for $f = f_n$, we get:

$$f = \frac{a}{1-b^2} \left(1 + b \frac{e^{i\Delta\theta} - b}{1 - be^{i\Delta\theta}} \right) = \frac{a}{1 - be^{i\Delta\theta}}$$

Therefore, the fixed points f further satisfy:[3]

$$\cos(\Delta\theta) = \frac{1}{2b} \left(1 + b^2 - \frac{a^2}{r^2} \right) \tag{8.17}$$

This equation can easily be solved provided that the angle $\Delta\theta$ is a function of r.

Be begin our analysis with the case where no area contraction takes place, so that $b = 1$. In this case the disk F is replaced by the entire plane. An interesting property of the system (8.5) in this case is that the resulting map often exhibits a symmetry along the line $x = \frac{a}{2}$, particularly when chaotic behaviour is present. This is different from the symmetry axis of the differential equation analogue where the symmetry axis is the line $x = 0$. In the difference equation case, the egg-shaped form is shifted to the right of the original positions of the system of coordinates.

As a possible explanation for this symmetry, let us consider the map in the form:

$$f_{n+1} = a + f_n e^{i\Delta\theta}$$

For any complex number z, it can be easily seen that its reflection with respect to the axis $x = \frac{a}{2}$ is $a - \bar{z}$, where $\bar{z}$ is as usual the complex conjugate of a complex number. If we denote the reflection of f_n by g_n, then we see that:

$$g_{n+1} = a - \bar{f}_{n+1} = a - \left(a + \bar{f}_n e^{-i\Delta\theta} \right) = (-a + g_n)e^{-i\Delta\theta}$$

---

[3]The derivation of this is also left as an exercise to the interested reader.

Multiplying through with $e^{i\Delta\theta}$, we see that the reflections satisfy a very similar recursive relation, except that it goes "backwards":[4]

$$g_n = a + g_{n+1}e^{i\Delta\theta}$$

Assuming that at some time the map passes very close to the $x = \frac{a}{2}$ line, the corresponding point will be very close to its reflection, and the evolution of the system from that point on will be very close to the reflection of the history of the system up to that point. This would cause the map to seem symmetric, especially when it exhibits chaotic behaviour. Similarly, equilibrium points will tend to exhibit symmetry around this axis.

We proceed now to discuss the equilibrium points of the system. We are therefore searching for the points where $x_{n+1} = x_n$ and $y_{n+1} = y_n$. This leads to $x = \frac{a}{2}$ and $y = \frac{a}{2}\cot\left(\frac{\Delta\theta}{2}\right)$. It is clear that there is no symmetry axis in the x direction, as y does not show stable magnitude since it is a function of the angle $\Delta\theta$. However, an approximation of y is achieved when the angle $\Delta\theta$ is small. Then

$$y \approx \frac{a}{\Delta\theta} = \frac{MG}{a^2}$$

The point $\left(x = \frac{a}{2}, y = \frac{MG}{a^2}\right)$ is therefore an equilibrium point. However, it is not a stable equilibrium point. The system may easily escape to infinity when y is higher than $\frac{MG}{a^2}$.

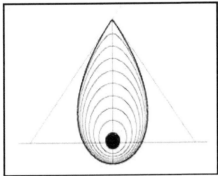

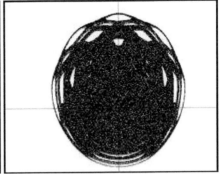

(a) Rotation-translation   (b) The central chaotic bulge

FIGURE 8.3:   Rotation-translation, and the chaotic bulge

---

[4] An important detail here is that $\Delta\theta$ depends on n, so the formulas don't quite match up, unless $\Delta\theta_{n+1}$ is close to $\Delta\theta_n$.

Figures 8.3(a) and 8.3(b) illustrate a rotation-translation case with parameters $a = 2$ and $MG = 100$. In Figure 8.3(a), a number of paths inside the egg-shaped formation are drawn. All the paths are analogous to those obtained by the differential equation analogue studied above, but they are moved to the right at a distance $x = \frac{a}{2}$. A chaotic attractor that resembles a bulge appears around the centre of coordinates. An enlargement of this attractor is illustrated in Figure 8.3(b). This is a symmetric chaotic formation with the line $x = \frac{a}{2}$ as an axis of symmetry.

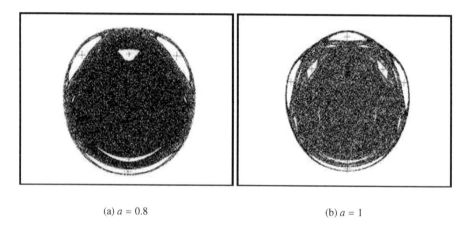

(a) $a = 0.8$ (b) $a = 1$

FIGURE 8.4:   The chaotic bulge

Figures 8.4(a) and 8.4(b) present an illustration of the chaotic bulge of the rotation-translation model. The parameters are $a = 0.8$ for Figure 8.4(a) and $a = 1$ for Figure 8.4(b). In both cases, $MG = 100$ and the rotation angle is $\Delta\theta = \sqrt{\frac{MG}{r^3}}$.

In Figure 8.4(b), third and fourth order equilibrium points appear, indicated by a small cross and a bigger cross respectively. In this case, the following relations hold: $x_{n+3} = x_n$ and $y_{n+3} = y_n$, and $x_{n+4} = x_n$ and $y_{n+4} = y_n$ respectively.

Accordingly, in Figure 8.4(a), two equilibrium cases appear, one of second order (indicated by small cross) where $x_{n+2} = x_n$ and $y_{n+2} = y_n$ and $y_{n+2} = y_n$ and one of the third order where $x_{n+3} = x_n$ and $y_{n+3} = y_n$.

In both cases, an algorithm is introduced for the estimation of the equilibrium points. The convergence is quite good provided that the starting values are in the vicinity of the equilibrium points.

As the parameter a takes higher values, higher order equilibrium points appear. A fifth order equilibrium point is added to those presented before, as illustrated in Figure 8.5(b), when $a = 1.2$. The relations for the new equilibrium point are $x_{n+5} = x_n$ and $y_{n+5} = y_n$. Figure 8.5(a) illustrates the case where $a = 0.2$. Only a first order equilibrium point is present.

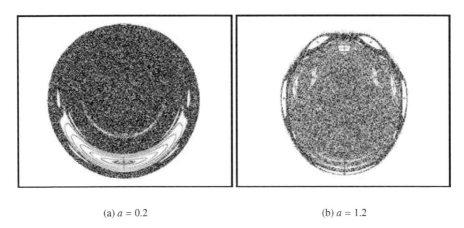

(a) $a = 0.2$ (b) $a = 1.2$

FIGURE 8.5:   The chaotic bulge

Higher order bifurcation is present in Figures 8.6(a) and 8.6(b). In Figure 8.6(a), a tenth order bifurcation form appears in the periphery of the chaotic attractor. The magnitude of the translation parameter is $a = 1.5$. In Figure 8.6(b), the translation parameter is $a = 1.9$. At this value, the bifurcation is so high that the chaotic region covers all the space bounded by the egg-shaped form. The point masses following chaotic trajectories eventually escape to infinity.

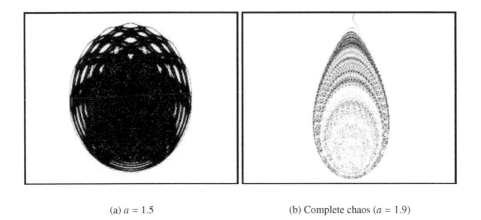

(a) $a = 1.5$ (b) Complete chaos ($a = 1.9$)

FIGURE 8.6:   The chaotic bulge

It is difficult to find the coordinates of equilibrium points that are located in the centre of the islands that are in the middle of the chaotic sea of the attractor, though as discussed earlier those are going to be symmetric with respect to the axis $x = \frac{a}{2}$.

A first order equilibrium point is characterised by the simple relations $x_{n+1} = x_n$ and $y_{n+1} = y_n$, which lead to the following relation for the radius of the circle on the periphery of which the equilibrium point must be located

$$R_k \approx \sqrt[3]{\frac{MG}{(2k\pi)^2}}, \quad k = 1, 2, \ldots$$

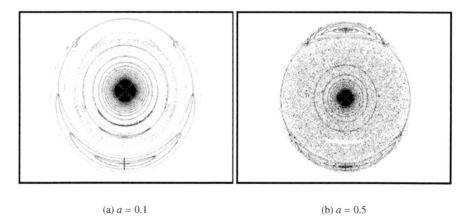

(a) $a = 0.1$ (b) $a = 0.5$

FIGURE 8.7: Equilibrium points

The circles R_k introduce a set of decreasing order with regards to the magnitude of the radius, as illustrated in Figure 8.7(a) ($a = 0.1$). The first point located on the circle with the larger radius ($k = 1$) is indicated by a cross. A few isoclines indicating the limits of the island are drawn around this point. Subsequent points follow in decreasing order. They are located in the middle of the isoclines.

The equilibrium points of higher order follow by using different relations. A simple case is that of the estimation of the equilibrium points of the second order, that is, when the coordinates are $x_{n+2} = x_n$ and $y_{n+2} = y_n$. This assumption leads to the following equation:

$$(x^2 + y^2)\cos^2(\Delta\theta) + 2x^2 \cos(\Delta\theta) + x^2 - y^2 = 0$$

This equation is fulfilled if the rotation angle obeys the relation $\Delta\theta = (2k + 1)\pi$. Then, this result leads to the following relation for the radius of the circle, on the periphery of which the equilibrium points must be located:

$$R_k \approx \sqrt[3]{\frac{MG}{((2k + 1)\pi)^2}}, \quad k = 1, 2, \ldots$$

The simulation results are illustrated in Figure 8.7(b). The translation parameter is $a = 0.5$. The concentric circles are distributed in decreasing order according to the magnitude of the radius R_k. The first two equilibrium points are distributed in the outer circle. They are indicated by a cross and are located on the axis of symmetry $x = \frac{a}{2}$, but in opposite directions: one to the positive and one to the negative part of the semi-plane for y. Three pairs of order two equilibrium points are illustrated in Figure 8.8(a). The equilibrium pairs are indicated by a cross. Also, the three equilibrium islands of the first order are presented. The parameter is $a = 0.04$. As this parameter has a very low value, the central chaotic bulge has a small radius.

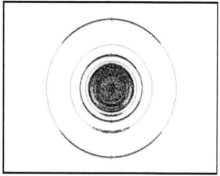 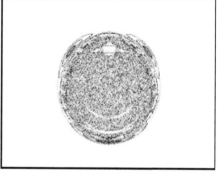

(a) Equilibrium pairs ($a = 0.04$)   (b) Order-3 bifurcation ($a = 0.7$)

FIGURE 8.8:   The chaotic bulge

The order three equilibrium points are presented in Figure 8.8(b). They are located in the middle of the three islands that are in the outer part of the chaotic bulge. The translation parameter is $a = 0.7$. The order three equilibrium points are in the periphery of a circle centred at $(\frac{a}{2}, \frac{a}{5})$. The radius of this circle is approximated by

$$R \approx \sqrt[3]{\frac{MG}{(2\pi/3)^2}} - \left(\frac{a}{4}\right)^2$$

8.4   A General Rotation-Translation Model

The solution of the differential equation of the model (8.7) leads to the form

$$ay + h = \int (\Delta\theta) r \mathrm{d}r$$

A simple approximation of the quantity $\Delta\theta$ is:

$$\Delta\theta \approx \sqrt{\frac{MG}{r^\beta}}$$

Then, the solution has the form

$$ay + h = \frac{r^{2-\frac{\beta}{2}}}{2-\frac{\beta}{2}}\sqrt{MG} \tag{8.18}$$

The special case $\beta = 2$ leads to the following equation form

$$ay + h = r\sqrt{MG}$$

or:

$$(x^2 + y^2)GM = (ay + h)^2$$

This equation expresses a family of ellipses (Figure 8.9(a)). The parameter a is 6, and the constant of integration h takes several values. The axis of symmetry is at $x = \frac{a}{2}$. Figure 8.9(b) illustrates a chaotic central bulge. The outer part of this chaotic attractor is approximated by an ellipse with $h = MG + a^2$.

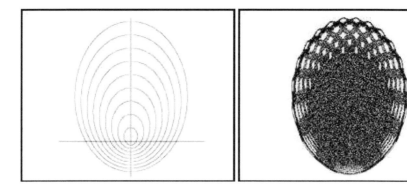

(a) A family of ellipses　　　　　　　　　(b) The chaotic bulge

FIGURE 8.9:　Elliptic forms and the chaotic bulge

When $\beta = 4$, the resulting equation for (x, y) is:

$$\sqrt{MG}\ln(r) = ay + h$$

The rotating forms are egg-shaped and similar to those provided in earlier rotation forms. With the exception of $\beta = 2$, where the rotation paths are elliptic, the cases with $\beta > 2$ provide egg-shaped rotation forms.

Another very important property of the elliptic case ($\beta = 2$) is that the rotating particles remain in the elliptic paths even if high values for the parameter h are selected. In the limit when the parameter a approaches zero the ellipses turn to concentric circles.

8.5 Rotating Particles inside the Egg-Shaped Form

Two cases are of particular importance. In the first case, the translation parameter a is quite small, and the particles are trapped inside the egg-shaped form and remain there, following the trajectories proposed by the theory discussed so far. However, a smaller region inside the trapping region exhibits chaotic behaviour. In this region, the particles follow chaotic paths that form the attractors presented above. In Figure 8.10(a), the outer limits of the egg-shaped form are drawn, and a disk of rotating particles of equal mass is centered at $(0,0)$. The particles are distributed by following the inverse law for the density ρ given by $\rho = \frac{c_i}{r^3}$. The diameter of the disk is chosen as to be exactly within the limits of the egg-shaped form. The parameters are $a = 0.25$ and $GM_0 = 0.45$. The rotation angle is given by

$$\Delta\theta = \sqrt{\frac{GM_0}{r^3}}$$

The resulting form that the disk of rotating particles takes after time $t = 10$ appears in Figure 8.10(b). The original cyclic form has now changed, providing an outer form of rotation and an inner chaotic attractor-like object.

Figures 8.10(c) and 8.10(d) illustrate the resulting picture after time $t = 20$ and $t = 100$ respectively. The distinct inner attractor is more clearly formed. In the case presented in Figure 8.10(d), the outer part of the rotating object extends to almost the entire space of the shape, but by following very specific characteristic paths.

In the second case, the translation parameter is high enough so that all of the egg-shaped space is contained in the chaotic region. The system is unstable and the rotating particles are not retained inside the egg-shaped region. On the contrary, they escape by following the escape trajectories and move away from the egg-shaped region. After a while, the majority of the particles will leave the region. Figures 8.11(a), 8.11(b) and 8.11(c) illustrate three instances, in times $t = 2$, $t = 3$ and $t = 4$ respectively. The translation parameter is $a = 0.6$. The cross in Figure 8.11(a) is at $(x, y) = (a, 0)$. A cross indicates the characteristic centres of the chaotic forms. These coordinates are calculated by using as starting values in a repeated procedure $x = a$ and $y = 0$. The iterative procedure is based on the difference equations for x and y.

The case when $t = 10$ appears in Figure 8.12. The rotating system of particles is in an intermediate stage. It escapes from the egg-shaped form, but a part of the system remains around the location $(x, y) = (a, 0)$ inside the egg-shaped form. When

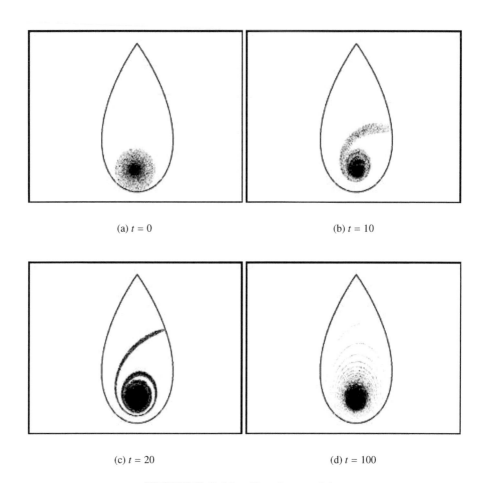

(a) $t = 0$ (b) $t = 10$

(c) $t = 20$ (d) $t = 100$

FIGURE 8.10:   Rotating particles

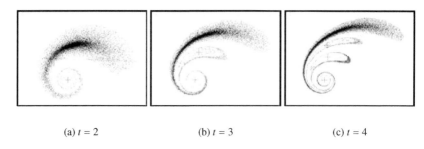

(a) $t = 2$ (b) $t = 3$ (c) $t = 4$

FIGURE 8.11:   Development of rotating particles over time ($a = 0.6$).

the magnitude of the translation parameter takes higher values, all the particles escape after a short interval (Figure 8.12(a)). An intermediate stage appears in the Figure 8.12(b), with parameter $a = 0.7$ and time $t = 10$. The leaf-like structure starts to disappear inside the egg-shaped form when close to $(x, y) = (a, 0)$. These jets eject material through the trajectories of the model. The speed of these jets depends on the magnitude of the parameter a and on the original rotation speed. The speed and the direction of the jets of material coincides to a after a large enough amount of time t. Figure 8.12(c) illustrates the case when $a = 0.8$. Only the last part of the jet remains connected to the source of the chaotic sea inside the egg-shaped pattern.

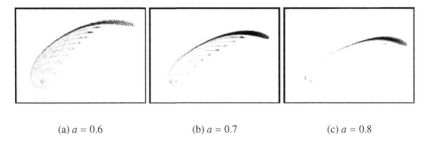

(a) $a = 0.6$ (b) $a = 0.7$ (c) $a = 0.8$

FIGURE 8.12:   Development of rotating particles according to the magnitude of the translation parameter. The elapsed time is $t = 10$.

When the translation parameter is $a = 1$ ($t = 10$), all the rotating material escapes outside the egg-shaped pattern. In Figure 8.13(a), the original cyclic disk is quite large, and a part of this disk lies outside the egg-shaped formation. However, all the material follows the escape route as a compact formation. On the other hand, in Figure 8.13(b), the influence of a small translation parameter ($a = 0.25$) to an original large cyclic cloud has a critical impact to the cloud formation. After $t = 100$, the cloud is separated into two formations, the first inside the egg-shaped form with the chaotic bulge-form in the middle, and the second outside.

8.6   Rotations Following an Inverse Square Law

We return to the original system of differential equations

$$\dot{x} = a - y\Delta\theta$$
$$\dot{y} = \quad x\Delta\theta$$

(8.19)

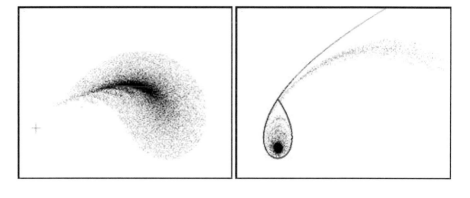

(a) $a = 1, t = 10$ (b) $a = 0.25, t = 100$

FIGURE 8.13:   Rotations

and the resulting differential equation for x, y:

$$\frac{dy}{dx} = \frac{x\Delta\theta}{a - y\Delta\theta}$$

This equation leads to the following form

$$a\,dy = (\Delta\theta)r\,dr$$

The solution of this differential equation depends on the form of the function $\Delta\theta$. In this section, we use a function known from mechanics, that deals with the rotation angle. This function arises from the law related to the transverse component of the acceleration in a circular movement. The function is expressed by the formula $\dot\theta = \frac{c_1}{r^2}$, where c_1 is a constant. Provided that $\Delta t = 1$, the following approximation for $\Delta\theta$ is obtained

$$\Delta\theta \approx \frac{c_1}{r^2}$$

Finally, we proceed to the solution of the equation:

$$a\,dy = \frac{c_1}{r}dr$$

The solution is

$$c_1 \ln(r) = ay + h \tag{8.20}$$

where h is an integration constant.

This last equation can be transformed in the following form:

$$\frac{c_1}{2} \ln(x^2 + y^2) = ay + h \tag{8.21}$$

Exploring the properties of this last equation, we set $x = 0$ when $y = r$. In this case the maximum or minimum values of y are obtained. The resulting equation for y is

$$c_1 \ln(y) = ay + h$$

The maximum value for y is achieved when $y = \frac{c_1}{a}$. Then, the value of the parameter h at this limit is

$$h_{\text{crit}} = c_1 \left(\ln \frac{c_1}{a} - 1 \right)$$

This value of h_{crit}, applied to equation (8.21), gives an equation for the path of a trajectory in the (x, y) plane, which divides the plane in two segments. This trajectory is the outer limit of the vortex region of the rotation. When $h > h_{\text{crit}}$, the trajectories diverge and the rotating object heads off to infinity. When $h < h_{\text{crit}}$, the object rotates inside the limits set by the above trajectory.

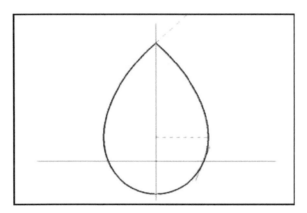

FIGURE 8.14: The characteristic trajectory

Some very interesting properties of the two-dimensional function are illustrated in Figure 8.14. The trajectory in the limit of escape is shaped as an egg. The sharper corner is at the maximum value of $y = \frac{c_1}{a}$ where $x = 0$. These values of x, y set the derivatives $\dot{x}, \dot{y}$ equal to zero. However, this point is not stable. This maximum value is obtained by solving the following equation for y:

$$\ln \left(\frac{ay}{c_1} \right) - \frac{ay}{c_1} + 1 = 0$$

The minimum value of y is obtained when the rotation speed is maximum. The resulting equation for y is:

$$\ln \left(\frac{ay}{c_1} \right) + \frac{ay}{c_1} + 1 = 0$$

A numerical solution gives $y = 0.278\frac{c_1}{a}$.

The maximum x_{max} is estimated by equating to zero the first derivative of the following equation for x, y:

$$\frac{c_1}{2} \ln(x^2 + y^2) = ay + c_1 \ln\left(\frac{c_1}{a}\right) - c_1$$

After appropriate differentiation, a numerical solution of the resulting equation gives $x_{max} = 0.402\frac{c_1}{a}$. This is achieved at $y = 0.203\frac{c_1}{a}$.

When $y = 0$, then $x = \frac{c_1}{ae}$. The tangent at this point is $\frac{dy}{dx} = e$. This tangent appears in Figure 8.14.

The tangent at the top sharp corner of the egg-shaped form can be computed as before, using Taylor approximation of the curves in a neighbourhood of the point $\left(0, \frac{c_1}{a}\right)$. Setting $z = y - \frac{c_1}{a}$, equation (8.21) becomes:

$$\frac{c_1}{a} \ln\left(x^2 + z^2 + 2\frac{c_1}{a}z + \frac{c_1^2}{a^2}\right) = az + c_1 + h \tag{8.22}$$

Using the expansion

$$\ln(a + x) = \ln(a) + \frac{x}{a} - \frac{1}{2}\frac{x^2}{a^2} + \cdots$$

where $a = \frac{c_1^2}{a^2}$ and $x = x^2 + z^2 + 2\frac{c_1}{a}z$, we see that equation (8.22) becomes:

$$az + c_1 + h = \frac{c_1}{2} \ln\left(\frac{c_1^2}{a^2}\right) + \frac{c_1}{2} \frac{a^2}{c_1^2}\left(x^2 + z^2 + 2\frac{c_1}{a}z\right) - \frac{c_1}{2} \frac{a^4}{2c_1^4}\left(x^2 + z^2 + 2\frac{c_1}{a}z\right)^2 + \cdots$$

Finally, after simplification, this becomes:

$$0 = \frac{c_1}{2}a^2c_1^2\left(x^2 - 2z^2 + \cdots\right)$$

In other words, the two tangent lines at the point $(x, z) = (0, 0)$ are given by the equation:

$$x^2 = 2z^2$$

In Figures 8.15(a) and 8.15(b), important properties of the model are presented. In Figure 8.15(a), the continuous model presented above is simulated. On the other hand, in Figure 8.15(b), the discrete analogue of the model appears. Figure 8.15(b) illustrates the central bulge in the case of $a = 2.7$ and $c_1 = 100$. The value of the translation parameter is quite high. This causes the appearance of an enormous chaotic attractor in the middle of the egg-shaped area. As in the case discussed previously, strong bifurcation is present.

As is illustrated in Figure 8.15(a), the particle follows an anti-clockwise direction. It starts from a point near zero, and it rotates away from the centre until the highest

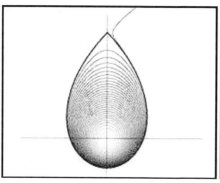

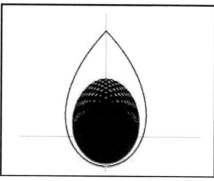

(a) The continuous case   (b) The central chaotic bulge

FIGURE 8.15: Comparison of the continuous and the discrete models

permitted place. Then, it escapes to infinity following the tangent on the top of the egg-shaped path. The iterative formula for the above simulations is the following:

$$x_{n+1} = x_n + \left(a - y_n \frac{c_1}{r^2}\right) d$$
$$y_{n+1} = y_n + x_n \frac{c_1}{r^2} d \tag{8.23}$$
$$r_n = \sqrt{x_n^2 + y_n^2}$$

The parameters in both figures are $a = 2.7$, $c_1 = 100$ and $d = 0.0001$.

Questions and Exercises

1. Show that the maximum speed in the rotating mass of system (8.7) is obtained exactly when
$$y = -\frac{MG}{a^2}, x = 0$$

2. Show that for $h > \frac{MG}{a}$ the orbits of system (8.7) go to infinity.

3. Consider the system (8.14), expressed via the simpler form (8.15). Show, by considering the resulting system for $z_n = \frac{f_n - a}{ab/(1-b)}$, that if f_n is within the circle centred at $(a, 0)$ and with radius $\frac{ab}{1-b}$, then f_{n+1} is also within that circle. What happens with point outside this circle?

4. Derive equation (8.17) for the fixed points of the system (8.14).

5. Show that if z is a complex number, its reflection along the line $x = \frac{a}{2}$ will be $a - \bar{z}$.

6. Construct the level curves in the x, y plane of equation (8.9) for $h = 0$, and varying values of a. What is the form of the curves when $a = 0$?

7. Use the equations (8.7) where
$$\Delta\theta = \sqrt{\frac{MG}{a^3}}$$
to determine the rotation speed in the x and y directions for the level curves examined in the previous question.

8. Construct the levels curves in the x, y plane of equation (8.9) when $h = \frac{2MG}{a}$, for various values of a.

9. Draw the level curves (x, y) of equation (8.18) for various values of β, and especially for $\beta = 1$.

10. Following the methodology of section 8.6, draw escape paths of particles through an egg-shaped form.

11. Construct a rotating model providing an egg-shaped form, and draw a chaotic bulge and escape paths for both small and large values of the parameter a.

12. Explain why the continuous case of Figure 8.15(a) does not show any chaotic behaviour.

13. Draw trajectories for the discrete case of Figure 8.15(a).

Chapter 9

Shape and Form

9.1 Introduction

About 2300 years ago, Euclid of Alexandria wrote his famous book on geometry called *The Elements*. This book, and its famous five postulates, became the basis of what is now known as Euclidean geometry. The usual notion of distance in space, essential when we design geometric forms in two or three dimensions, carries his name (the "Euclidean metric").

The development of science throughout the centuries brought new theories and improvements in all scientific fields. Nevertheless, the Euclidean metric remains unchanged, and still plays an essential part in a bevy of topics. The theory of the simulation of chaotic shapes included in this book is based on an analytic approach that for the most part uses the fundamental notions of traditional Euclidean geometry. The essential components of this theory, namely the notions of *translation*, *rotation* and *reflection*, are defined and explored in this chapter, along with applications related to the development of chaotic forms and shapes.

As one can readily verify, a large number of chaotic objects that appear in the literature may be classified as geometric objects produced by following simple geometric rules. Even non-chaotic objects are similarly simulated by following the same geometric rules.

Translation, rotation and reflection are not only tools essential in forming geometric objects, but can also be used as a basis of understanding how a process, chaotic or not, can generate *shapes* and *forms*. The linear movements of physical objects are easily modelled geometrically by using translation. Rotation and translation are present in stellar systems, galaxies, tornados and vortex movements. Reflection is generally associated with light refraction, galactic object formation, as well as vortex formation.

This chapter includes a brief introduction to the elementary rules of analytical geometry that are used throughout this book, along with several applications on the shape and form of chaotic and non-chaotic objects. This theoretical treatment is combined with applications to special relativity and Penrose's tiling theory, via simulations based on relatively simple affine transformations. Interested readers may find related topics in books on hyperbolic geometry.

9.1.1 Symmetry and plane isometries

Any subset of the plane is called a *figure*. It can be easily observed that some figures are more interesting than others. Figures with a high degree of symmetry are most interesting, not only because of interesting forms arising in simulations, but also because they often occur in nature. Spiral galaxies, snowflakes, molecules and crystals are examples of objects with symmetric cross sections.

By *symmetry* for a figure A, we mean simply a transformation that preserves the figure. The simplest kind of symmetry in a plane figure is that of symmetry with respect to a line. Two points X and X' are *symmetrical with respect to a line l*, if the line l bisects and is perpendicular to the segment XX'. We will also say that X' is the *reflection* of X along ℓ (Figure 9.1).

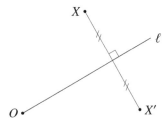

FIGURE 9.1: Symmetry/reflection along a line

An *isometry* is a transformation which preserves distances; *i.e.* it is a transformation

$$T : \mathbb{R}^2 \to \mathbb{R}^2$$

with the property that the distance from $T(A)$ to $T(B)$ is the same as the distance from A to B. A standard theorem in Euclidean geometry is that

> every planar isometry is completely determined by its effect on three non-collinear points.

The proof is fairly straightforward and instructive: Since distances are preserved, the circle centred at A and with a given radius is mapped to the circle centred at TA, and with the same radius. If we consider the three circles through D, and with centres A, B, C respectively, and look at their images, then TD must be the point of intersection of these three circles (Figure 9.2).

Isometries respect the two fundamental aspects of geometry: the incidence aspect based on the notion of *collinearity*, and the metric aspect based on the notion of *distance*. The fact that isometries respect collinearity is left as an exercise.

There are four basic kinds of planar isometries:

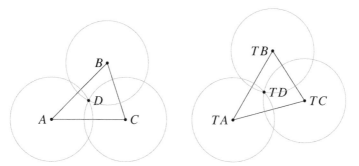

FIGURE 9.2: Planar isometries are determined by their effect on three points

Translations These are fully described by a single vector $\vec{v}$. Such a transformation is given simply by:[1]

$$T\vec{w} = \vec{w} + \vec{v}$$

Rotations These correspond to a rotation with a given angle and around a specific point. Assuming the rotation is around the origin, the transformation is given simply by the matrix:

$$\mathbf{rot}\,\theta = \begin{bmatrix} \cos\theta & -\sin\theta \\ \sin\theta & \cos\theta \end{bmatrix}$$

in other words:

$$T\vec{w} = \mathbf{rot}(\theta)\begin{bmatrix} x \\ y \end{bmatrix} = \begin{bmatrix} x\cos\theta - y\sin\theta \\ x\sin\theta + y\cos\theta \end{bmatrix}$$

Reflections A reflection across a line ℓ passing through the origin and forming angle θ with the x axis is given by the matrix

$$\mathbf{ref}\,\theta = \begin{bmatrix} \cos 2\theta & \sin 2\theta \\ \sin 2\theta & -\cos 2\theta \end{bmatrix}$$

and therefore it has the form:

$$T\vec{w} = \mathbf{ref}(\theta)\begin{bmatrix} x \\ y \end{bmatrix} = \begin{bmatrix} x\cos 2\theta + y\sin 2\theta \\ x\sin 2\theta - y\cos 2\theta \end{bmatrix}$$

Glide Reflections A glide reflection is a translation followed by a rotation along a line parallel to the translation direction. The translation and rotation can be performed in either order. An example of a glide reflection is given in figures 9.3(a) and 9.3(b). You can see that a glide reflection can be thought of also as the composition of a rotation followed by a reflection (or equivalently, three reflections along lines that do not all have a point in common).

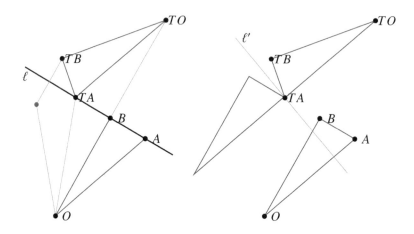

(a) Glide reflection as a rotation, followed by a reflection

(b) Glide reflection as a translation, followed by a reflection

FIGURE 9.3:   Glide reflection seen in two different ways

The matrices **rot** θ, **ref** θ have a series of useful properties, which we summarise here:

$$\mathbf{rot}\theta\,\mathbf{rot}\phi = \mathbf{rot}(\theta + \phi) \tag{9.1}$$

$$\mathbf{rot}(0) = \mathbf{I} \tag{9.2}$$

$$\mathbf{ref}\theta\,\mathbf{ref}\phi = \mathbf{rot}(2(\theta - \phi)) \tag{9.3}$$

$$\mathbf{ref}\theta\,\mathbf{rot}\phi = \mathbf{ref}(\theta - \phi/2) \tag{9.4}$$

$$\mathbf{rot}\phi\,\mathbf{ref}\theta = \mathbf{ref}(\theta + \phi/2) \tag{9.5}$$

$$\mathbf{ref}\theta\,\mathbf{ref}\phi\,\mathbf{refy} = \mathbf{ref}(\theta - \phi + \mathbf{y}) \tag{9.6}$$

We leave the verification of these identities to the reader.

It is worth pointing out that the composition of two reflections with the same centre is a rotation with angle twice the angle between the two reflection lines:

$$\mathbf{ref}\,\theta\,\mathbf{ref}\,\phi = \begin{bmatrix} \cos 2(\theta - \phi) & -\sin 2(\theta - \phi) \\ \sin 2(\theta - \phi) & \cos 2(\theta - \phi) \end{bmatrix} = \mathbf{rot}\,2(\theta - \phi)$$

For a geometric proof of this fact, consider the effect of the combined transformation on the points O, A, B in Figure 9.4(a). The first reflection, along the line ℓ_1, takes the points O, A, B to the points O, A, B',[2] and the second, along the line ℓ_2, takes those

---

[1] Here, as well as in the rest, we will identify a point $P(x, y)$ with the vector $\vec{w}$ starting at the origin and ending at (x, y).

[2] Since a reflection fixes all points on the reflection line.

points to the points O, A', B'. The same effect would be obtained by the aforementioned rotation.

At this point, you might be wondering, what happens when we take the composition of two reflections along parallel lines. The result is a translation with a vector twice the distance between the two lines (Figure 9.4(b)).

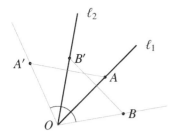

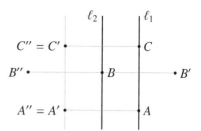

(a) The composition of two reflections along intersecting lines is a rotation around the intersection

(b) The composition of two reflections along parallel lines is a translation

FIGURE 9.4:   The composition of two reflections

A very striking result is that these four types, along with the identity, are all the plane isometries, namely:

1. Every plane isometry is a composition of reflections.

2. Every composition of reflections in the plane can be performed with at most three reflections.

The relation between isometry groups and symmetries is given by the following theorem, already known to Leonardo da Vinci:

All finite isometry groups can occur as symmetry groups of figures, and cannot contain translations or glide reflections.

We finish this section with a brief discussion of the points and shapes left invariant by the various isometries. A shape/set S is said to be *left invariant* by a transformation T, if the sets $T(S)$ and S coincide (*i.e.* if the transformation preserves S as a whole, though it might move the points of S around). We omit the identity, which leaves everything fixed.

Points Translations and glide reflections have no fixed points. A rotation fixes its centre, while a reflection fixes every point on the line of reflection.

Lines A translation along a line ℓ leaves that line, as well as any line parallel to it, invariant. A glide reflection leaves its line of reflection invariant. A rotation has no fixed lines, unless it is a rotation by 180°, in which case it leaves any line passing through the centre of rotation invariant. Finally, a reflection leaves invariant the line of reflection, as well as any line perpendicular to it.

9.2   Isometries in Modelling

We now focus our attention on maps that make heavy use of isometries in their construction.

9.2.1   Two-dimensional rotation

Rotation in a plane is the most frequently arising case in chaotic modelling. Even the three dimensional case of rotation, in many cases, is reduced to rotation in the (x, y) plane around an axis of rotation perpendicular to this plane, after an appropriate change of coordinates.

A map arising from rotation in the plane is expressed by:

$$\begin{bmatrix} x_{n+1} \\ y_{n+1} \end{bmatrix} = \mathbf{rot}\,\theta_n \begin{bmatrix} x_n \\ y_n \end{bmatrix} = \begin{bmatrix} x_n \cos\theta_n - y_n \sin\theta_n \\ x_n \sin\theta_n + y_n \cos\theta_n \end{bmatrix} \tag{9.7}$$

Orbits of the above map stay in a circle of radius $r = \sqrt{x_0^2 + y_0^2}$, where (x_0, y_0) is the set of initial values. This can be easily verified from the relation $r_{n+1} = r_n$, and is independent of the form of the rotation angle θ_n. Every system of this type will continue with circular movements without any change in the circular path.

When a translation is added to rotation, things change radically. Without loss of generality, we will consider the translation as a single parameter a added in the x direction. The resulting map is given by:

$$\begin{bmatrix} x_{n+1} \\ y_{n+1} \end{bmatrix} = \mathbf{rot}\,\theta_n \begin{bmatrix} x_n \\ y_n \end{bmatrix} + \begin{bmatrix} a \\ 0 \end{bmatrix} = \begin{bmatrix} a + x\cos\theta_n - y\sin\theta_n \\ x\sin\theta_n + y\cos\theta_n \end{bmatrix} \tag{9.8}$$

By the addition of a translation parameter, the radius r_{n+1} is not generally equal to its previous value r_n, as is easily verified by the relation:

$$r_{n+1}^2 = r_n^2 + a^2 + 2a(x_n \cos\theta_n - y_n \sin\theta_n)$$

It is clear that r_{n+1} is also a function of the rotation angle θ_n. Also, another relation, not directly containing the rotation angle, holds:

$$r_{n+1}^2 - 2ax_{n+1} + a^2 = r_n^2$$

This relation, after summation, yields

$$\sum_{1}^{n}(r_{i+1}^2 - r_i^2) + na^2 = 2a\sum_{1}^{n}x_i$$

where $i = 1, 2, \ldots, n$. Setting $\bar{x} = \frac{1}{n}\sum_{1}^{n}x_i$, and after the appropriate cancellations, we obtain:

$$2a\bar{x} = \frac{(r_n^2 - r_0^2)}{n} + a^2$$

In the limit ($n \to \infty$), assuming the radius r_n is bounded, the following simple relation arises:

$$\bar{x} = \frac{a}{2}$$

This is suggesting the possibility of symmetry around this axis in a rotation-translation iterative procedure (see for example Figure 9.5). This subject was discussed more extensively in Chapter 8.

One can easily obtain various symmetric forms by choosing a function for the rotation angle. Selecting non-linear functions for the rotation angle is essential for producing forms of high complexity, as well as chaotic forms. The figures in the subsequent sections are chaotic and non-chaotic cases of rotation-translation iterative schemes.

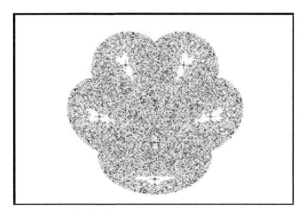

FIGURE 9.5: A symmetric graph

9.3   Reflection and Translation

Movement that follows reflection with translation is modelled as in the previous rotation-translation case. The map that expresses translation-reflection has the general form:

$$\begin{bmatrix} x_{n+1} \\ y_{n+1} \end{bmatrix} = \mathbf{ref}\,\theta_n \begin{bmatrix} x_n \\ y_n \end{bmatrix} + \begin{bmatrix} a \\ 0 \end{bmatrix} = \begin{bmatrix} a + x\cos 2\theta_n + y\sin 2\theta_n \\ x\sin 2\theta_n - y\cos 2\theta_n \end{bmatrix} \tag{9.9}$$

It is easily verified that the reflection-translation map (9.9) provides symmetric iterative forms with a symmetry axis perpendicular to the x axis at a distance equal to $a/2$ from the origin. Some illustrative examples follow.

9.3.1   Space contraction

The iterative procedures presented above using rotation, reflection and translation are based on a stable distance and space metric. At a first glance, space contraction or expansion could be a case of specific interest in relativistic problems in cosmology and elsewhere. However, several other very interesting cases appear. Many chaotic objects are formed by applying space contraction rules along with rotation, translation and reflection.

Space contraction or expansion appears in vortex formation in chaotic advection or in chemical reactions. The simplest model uses a parameter b by which the distance equations are multiplied. In other words, in every iteration we would have:

$$r_{n+1} = br_n$$

In the case of rotation-translation, the iterative formula takes the form

$$\begin{bmatrix} x_{n+1} \\ y_{n+1} \end{bmatrix} = b\,\mathbf{rot}\,\theta_n \begin{bmatrix} x_n \\ y_n \end{bmatrix} + \begin{bmatrix} a \\ 0 \end{bmatrix} = \begin{bmatrix} a + bx_n\cos\theta_n - by_n\sin\theta_n \\ bx_n\sin\theta_n + by_n\cos\theta_n \end{bmatrix} \tag{9.10}$$

The system (9.10) has a Jacobian determinant of $J = b^2$, which means a space contraction or expansion proportional to b^2. Space expansion is present for example in supernova explosions, but also in weather modelling and wave propagation. On the other hand, space contraction is more interesting from the point of view of stability and form generation. Contraction is counterbalanced by the presence of either rotation or translation. In the case of contraction, the iterative spiraling sink towards a point of attraction is replaced, due to rotation and other balancing mechanisms, to a movement inside an attracting space, the "*attractor.*"

It is interesting that the relativistic space contraction may lead to the same iterative scheme as (9.10). This is easily demonstrated by observing the relativistic space equations, that is

$$x_{t+1} = x_t\sqrt{1 - \frac{v^2}{c^2}} = bx_t$$

$$y_{t+1} = y_t\sqrt{1 - \frac{v^2}{c^2}} = by_t \tag{9.11}$$

where

$$b = \sqrt{1 - \frac{v^2}{c^2}}$$

It is evident in the previous formulae, that relativistic space contraction can be observed for relatively high speeds. However, as demonstrated in Chapter 12, changes can appear in chaotic images even in lower speeds.

9.4 Application in the Ikeda Attractor

A very interesting approach regarding rotation and reflection along with translation and space contraction appears in the figures below. The so-called *Ikeda model* and the famous associated attractor express the rotation-translation-space contraction example (Ikeda, 1979; Ikeda et al., 1980). The iterative scheme for rotation is:

$$\begin{bmatrix} x_{n+1} \\ y_{n+1} \end{bmatrix} = b \, \mathbf{rot} \, \theta_n \begin{bmatrix} x_n \\ y_n \end{bmatrix} + \begin{bmatrix} a \\ 0 \end{bmatrix}$$

Ikeda proposed the following function for the rotation angle:

$$\theta_n = c - \frac{d}{1 + x_n^2 + y_n^2} \tag{9.12}$$

where c and d are rotation parameters. The rotation angle θ is a function of the square of the distance from the origin. At large distances, the system rotates with an angle $\theta \approx c$, whereas, close to the origin, the rotation angle is $\theta \approx c - d$. The classical Ikeda attractor is obtained when $a = 1, b = 0.83, c = 0.4$ and $d = 6$ (Figure 9.6).

FIGURE 9.6: The Ikeda attractor

The reflection-translation map with a similar function to (9.12) for the reflection angle is expressed by the iterative scheme:

$$\begin{bmatrix} x_{n+1} \\ y_{n+1} \end{bmatrix} = b \, \mathbf{ref} \, \theta_n \begin{bmatrix} x_n \\ y_n \end{bmatrix} + \begin{bmatrix} a \\ 0 \end{bmatrix}$$

where the reflection angle is given by:

$$2\theta_n = c + \frac{d}{1 + x_n^2 + y_n^2}$$

The Jacobian determinant in this case is negative, $J = -b^2$. The reflection attractor, presented in Figure 9.7, has two separate images. The figure on the right looks like a mirror image of the Ikeda attractor. The parameters for this simulation were $a = 2.2$, $b = 0.9$, $c = 1.5$ and $d = 2.15$.

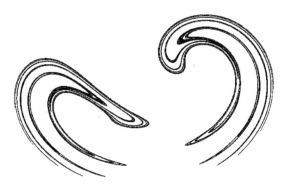

FIGURE 9.7:   The reflection Ikeda attractor

9.5   Chaotic Attractors and Rotation-Reflection

Both rotation and reflection generate chaotic attractors for appropriate values of the parameters a, b, c, and d. Two important questions are:

a) Are there parameter values that generate chaotic attractors in both cases?

b) If there are, then what are the interrelations between rotation and reflection attractors?

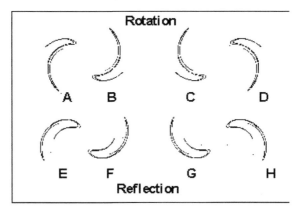

FIGURE 9.8: Comparing rotation-reflection cases

In Figure 9.8, we see four combinations for rotation, $A(\alpha, \theta)$, $B(-\alpha, \theta)$, $C(\alpha, -\theta)$ and $D(-\alpha, -\theta)$, compared to the four cases for reflection, $E(\alpha, \theta)$, $F(-\alpha, \theta)$, $G(-\alpha, \theta)$ and $H(-\alpha, -\theta)$. A rotation-reflection iterative scheme was used, just as in the preceding section, but with a rotation-reflection angle given by the more simple formula

$$\theta_n = c \sqrt{x_n^2 + y_n^2}.$$

In reflection, the parameter $2\theta_n$ was set by the same formula. The parameters were $a = 1$, $b = 0.68$ and $c = 10/3$. Figure 9.9 illustrates the first two graphs to the left of Figure 9.8.

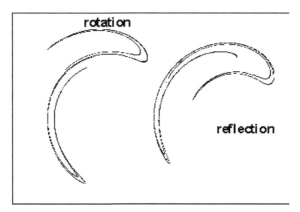

FIGURE 9.9: Comparing rotation-reflection case

Another reflection-translation chaotic attractor is formed by selecting a simple quadratic function for the reflection parameter θ, that is

$$2\theta_n = c\sqrt{x_n^2 + y_n^2}.$$

The parameters for the simulation in Figure 9.10 were $a = 1$, $b = 0.68$ and $c = 10/3$.

FIGURE 9.10:   A reflection chaotic attractor resulting from a quadratic rotation angle

The same reflection model provides two other attractors, as illustrated in Figure 9.11. Both figures are obtained for $a = 1$ and $b = 0.68$, whereas the other parameter is $c = 1$ for the image on the left and $c = 2$ for the image on the right.

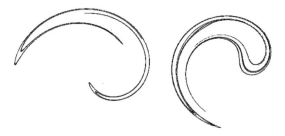

FIGURE 9.11:   Two other reflection attractors from a quadratic rotation angle

9.6   Experimenting with Rotation and Reflection

9.6.1   The effect of space contraction on rotation-translation

In order to explore the effect of space contraction due to the introduction of a parameter b ($b = 0.8 < 1$), consider the following rotation-translation scheme:

$$x_{n+1} = a + b(x_n \cos \theta_n - y_n \sin \theta_n)$$
$$y_{n+1} = \quad\quad b(x_n \sin \theta_n + y_n \cos \theta_n)$$

(9.13)

where the rotation angle is

$$\theta_n = c + \frac{d}{x_n^2 + y_n^2}$$

The Jacobian determinant in this case is $J = b^2 = 0.64$. The translation parameter is $a = 6$, the space contraction parameter is $b = 0.8$ and the rotation parameters are $c = a/2$ and $d = a$. The original particles are located in a circle of radius 6 centred at $(x, y) = (a/2, 0)$. The particles are attracted into the two chaotic attractors (Figure 9.12(b)) inside the circle. The time development of the process is illustrated in Figure 9.12(a). The outer circle represents the original particles at time $t = 0$. At time $t = 1$, the particles form an ellipsoid form, whereas at time $t = 2$, the outer part of the first chaotic attractor appears. According to this computer experiment the approach to the chaotic state is quite rapid. However, the parameters of the model used are high and the model expresses a strong chaotic process.

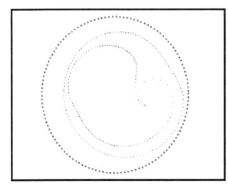 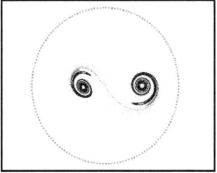

(a) The space contraction effect   (b) Space contraction and chaotic images

FIGURE 9.12:   Space contraction

9.6.2 The effect of space contraction and change of reflection angle on translation-reflection

A translation-reflection model is introduced and applied in the following computer simulation. The set of equations expressing this model are

$$x_{n+1} = a + b(x_n \cos \theta_n + y_n \sin \theta_n)$$
$$y_{n+1} = \quad b(x_n \sin \theta_n - y_n \cos \theta_n)$$

(9.14)

where the rotation angle is:

$$\theta_n = c + \frac{d}{1 + x_n^2 + y_n^2}$$

The original set of parameters is $a = 2.2$, $b = 0.9$, $c = 3$. The parameter d varies, taking the following values: 32, 16, 8, 6, 4.78 and 4.30 respectively. Six chaotic forms of the reflection-translation scheme are presented here. The original form appears on the top left of Figure 9.13, and the other images follow from left to right in the top row and from left to right in the second row. We see that, as the parameter d decreases, the original object is gradually divided into two distinct objects.

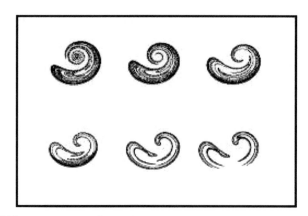

FIGURE 9.13: The effect of the reflection parameter on chaotic images

The influence that a change in the space parameter b has on a chaotic system is examined in the following reflection-translation scheme. The original object results when $a = 2.2$, $c = 3$, $d = 4.3$ and $b = 1$, whereas for the other five objects the space-contracting parameter decreases, taking the values 0.99, 0.975, 0.94, 0.93 and 0.9 respectively. The original form appears on the top left of Figure 9.14, and the other images follow from left to right in the top row and from left to right in the second row. Once again, as b decreases, the original object is divided into two distinct objects, but in this case following a process different from that illustrated in Figure 9.13.

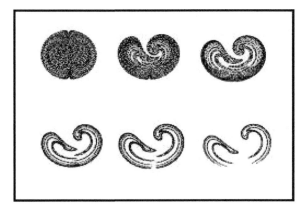

FIGURE 9.14: The effect of space contraction on chaotic images

9.6.3 Complicated rotation angle forms

More complicated chaotic forms arise by selecting a rotation-translation scheme with rotation angle that follows a more complicated equation. In this case, the rotation angle is a function of the inverse of the radius $r = \sqrt{x^2 + y^2}$ plus a distance parameter c and another rotation angle parameter m. The area contraction parameter is set to $b = 0.93$. The function for the rotation angle is

$$\theta_n = -\frac{1}{\sqrt{c^2 + x^2 + y^2}} - \frac{m}{\sqrt{c^2 + x^2 + dy^2}}$$

The parameters for the simulation were set to $a = 0.74$, $c = 0.5$, $d = 1.6$ and $m = 3.1$. The resulting chaotic attractor is illustrated in Figure 9.15. This is a more complicated chaotic attractor that could result in chaotic flows and in galactic formations.

FIGURE 9.15: The effect of a complicated rotation angle on chaotic attractors

A simpler form is produced by assuming that in the function for the rotation angle the parameter c is 0. The resulting chaotic system provides an attractor expressing flows appearing in the mixing of fluids but, also, in other rotation-translation systems. The parameters in Figure 9.16 were $a = 1$, $b = 0.87$, $d = 1$ and $m = 3.1$.

FIGURE 9.16:   The effect of a complicated rotation angle on chaotic attractors

A two-piece chaotic attractor is formed by using a rotation-translation formula with a special rotation angle. The rotation angle has a constant term denoted by c and a varying term related to a transformed distance. The model is a space contracting one ($b = 0.9$). It also has a slightly different form for the rotation-translation

$$
\begin{aligned}
x_{n+1} &= a + b(x_n \cos \theta_n - y_n \sin \theta_n) \\
y_{n+1} &= -a + b(x_n \sin \theta_n + y_n \cos \theta_n)
\end{aligned}
\tag{9.15}
$$

where the rotation angle is

$$
\theta_n = c + \frac{d}{\left(\frac{x-a}{b}\right)^2 + \left(\frac{y+a}{b}\right)^2}
$$

The other parameters were set to $a = 9$, $c = a$ and $d = 6$. The two-piece chaotic attractor is presented in Figure 9.17(a).

9.6.4   Comparing rotation-reflection

Figure 9.17(b) illustrates a two-piece chaotic attractor, resulting from a set of equations expressing rotation and translation without space contraction ($b = 1$). The rotation angle is a function of r, that is

$$
\theta_n = c + \frac{d}{\sqrt{x_n^2 + y_n^2}}.
$$

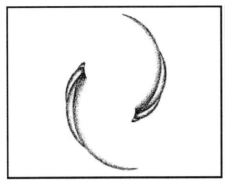 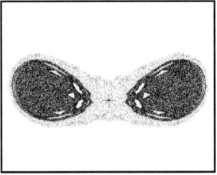

(a) With space contraction   (b) Without space contraction

FIGURE 9.17: Two-piece chaotic attractors

The process shows how an original rotating disk of particles can form two distinct symmetric chaotic forms. The parameters were set to $a = 37$, $c = 2.85$ and $d = 15.5$.

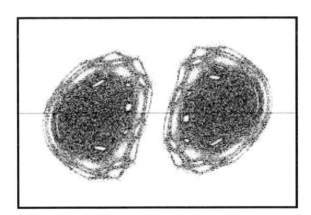

FIGURE 9.18: Reflection: a two-piece chaotic attractor

The images presented in Figure 9.18 are the reflection analogue of this system. The reflection-translation equations without space contraction ($b = 1$) are used. The reflection angle (θ_{ref}) is as in the previous rotation-translation case but it is related to the rotation angle (θ_{rot}) by the relation $\theta_{ref} = 2\theta_{rot}$. When the parameters take the values $a = 80$, $c = 5$ and $d = -20$, the reflection-translation model provides two objects with mirror symmetry, as illustrated in Figure 9.18.

9.6.5   A simple rotation-translation model

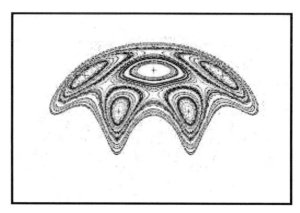

FIGURE 9.19:   A simple area preserving rotation-translation model

A simple space preserving ($b = 1$) rotation-translation model with a rotation angle, varying according to the square of the distance r from the origin ($\theta_n = c(x_n^2 + y_n^2) = cr^2$), also gives an interesting rotation-translation image, illustrated in Figure 9.19. There is an island with equilibrium points achieved when ($x_{n+1} = x_n, y_{n+1} = y_n$), centred on the symmetry axis ($x = a/2$), and four symmetric islands (in two pairs) with equilibrium points that follow the rule $x_{n+4} = x_n$ and $y_{n+4} = y_n$. The parameters were set to $a = 1$ and $c = 0.65$.

9.7   Chaotic Circular Forms

The study of a rotation process in which the rotating particles move at circular or other paths from the centre following a specific formula is of particular interest. The simulations arising out of these cases start by forming concentric circles with different particle density and can lead to more complicated forms, depending on the selected set of non-linear functions. The iterations follow the simple scheme:

$$
\begin{aligned}
x_n &= r_n \cos \theta_n \\
y_n &= r_n \sin \theta_n
\end{aligned}
\tag{9.16}
$$

where θ_n increases in a constant way:

$$\theta_{n+1} = \theta_n + \theta$$

Depending on the iterative scheme for the radius r_n, interesting patterns arise, along with chaotic behaviour. As a simple example, consider the iterative formula:

$$r_{n+1} = b - \frac{a}{|r_n|}$$

The (a, r) bifurcation diagram for this model is shown in Figure 9.20(a). Second, third, fourth and higher order bifurcations appear. The parameter b is set to 2.4, while the parameter a varies in $1.2 < a < 6.9$.

Figure 9.20(b) illustrates the bifurcation diagram (a, r) of the model expressed by the formula:

$$r_{n+1} = b - \frac{a}{r_n^2}$$

The parameter b is set to 2.4, while the parameter a varies in $0.2 < a < 60$.

Figure 9.20(c) illustrates the (a, r) bifurcation diagram of the model:

$$r_{n+1} = b - \frac{a}{r_n^4}$$

The parameter b is set to 2.4 and the parameter a varies in $0.2 < a < 60$.

The development of the theory of rotation with varying rotation distance from the origin, or with varying rotation angle related to the distance $r = \sqrt{x^2 + y^2}$, is partly explained by using Figure 9.21(a), where $OC = r$.

A very interesting rotation scheme is presented in Figure 9.21(b). There is only one parameter $b = 1.195$. The radius r follows an inverse square law:

$$r_{n+1} = \frac{1}{(r_n - b)^2} \tag{9.17}$$

Chaotic behaviour appears when the parameter b varies. A small change in b leads to the onset of chaos. For $b = 1.204$, the bifurcation has already started to divide the outer ring in two rings (Figure 9.21(c)). The bifurcation process leads to chaos when higher values for b are selected.

In Figure 9.21(d), there appear the distinct circles. By giving to the chaotic parameter b the value 1.255, the particles are distributed in a particular chaotic way. There are cyclical regions with higher or lower densities, as presented in Figure 9.21(e).

Figure 9.21(f) presents the bifurcation diagram (b, r) for the model (9.17). The parameter b varies between $0 < b < 2.2$. The model gives two values for r in the region of $0 < b < 0.629 \cdots$. At $0.629 \cdots$, as b increases, a second bifurcation starts, leading to a chaotic region. At the end of this region, three values for r appear, making up a new chaotic region, then four values, and so on until a totally chaotic region at $1.88 \cdots$.

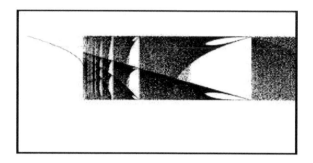

(a) $r_{n+1} = b - \frac{a}{|r_n|}$

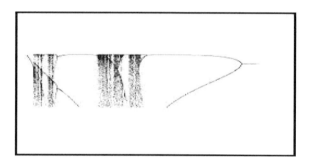

(b) $r_{n+1} = b - \frac{a}{r_n^2}$

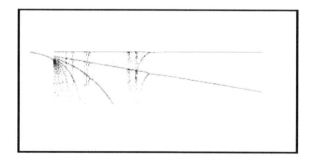

(c) $r_{n+1} = b - \frac{a}{r_n^4}$

FIGURE 9.20: Bifurcation diagrams

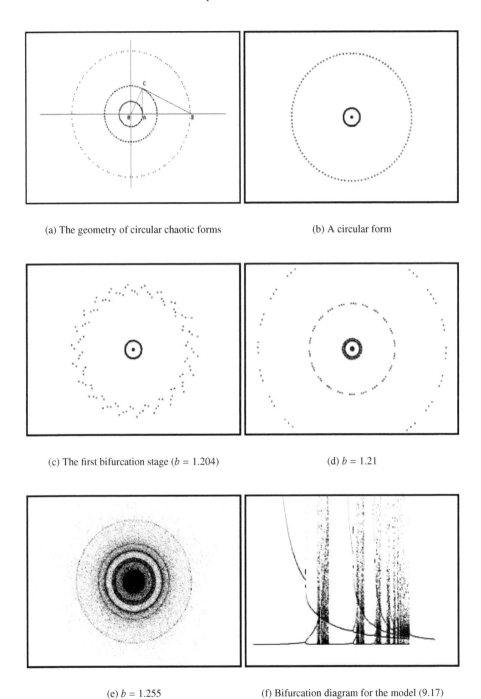

(a) The geometry of circular chaotic forms

(b) A circular form

(c) The first bifurcation stage ($b = 1.204$)

(d) $b = 1.21$

(e) $b = 1.255$

(f) Bifurcation diagram for the model (9.17)

FIGURE 9.21: Circular chaotic forms

9.8   Further Analysis

Consider a rotation-translation case expressed by the relations

$$x_{n+1} = a + x_n \cos \theta_n - y_n \sin \theta_n$$
$$y_{n+1} = \quad\; x_n \sin \theta_n + y_n \cos \theta_n \qquad (9.18)$$

or in the equivalent form

$$z_{n+1} = a + z_n e^{i\theta_n}$$

where $z_n = x_n + iy_n$.

This map is symmetric with the line $x = a/2$ as the axis of symmetry. This is easily demonstrated if we consider the fixed points, *i.e.* solutions of the form $z_{n+1} = z_n = z$. In other words, when $(x_{n+1}, y_{n+1}) = (x_n, y_n) = (x, y)$. Now the following relation holds:

$$|z - a| = |z|$$

or

$$(x - a)^2 + y^2 = x^2 + y^2$$

and, finally,

$$x = \frac{a}{2}$$

whereas for y

$$y = \frac{a}{2} \cot \frac{\theta}{2}$$

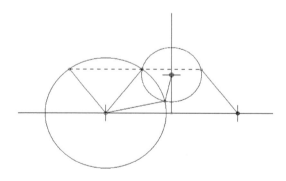

FIGURE 9.22:   Geometric analogue of rotation-translation ($\theta = 2$)

A geometric analogue of this rotation-translation case is illustrated in Figure 9.22 where the translation parameter is $a = 2$, the rotation angle is $\theta = 2$ and the initial values are $(x, y) = (0.9, 0.2)$. Figure 9.23(a) illustrates the same case, but now the rotation angle is $\theta = \pi$. In this case the point $(x, y) = (a/2, 0)$ indicated by two concentric small circles is the equilibrium point. An equilibrium point located at $(x, y) = (a/2, a/2)$ is presented in Figure 9.23(b) where the value of the rotation angle is $\theta = \pi/2$.

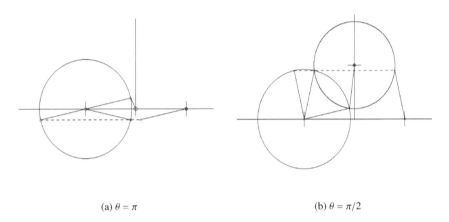

(a) $\theta = \pi$ (b) $\theta = \pi/2$

FIGURE 9.23: Geometric analogues of rotation-translation

The Jacobian determinant for the rotation-translation case above is

$$J = 1 + x_n \frac{\partial \theta_n}{\partial y} - y_n \frac{\partial \theta_n}{\partial x} = 1 + x_n \theta_y - y_n \theta_x$$

where

$$\theta_x = \frac{\partial \theta_n}{\partial x}$$
$$\theta_y = \frac{\partial \theta_n}{\partial y}$$

Analogously, the Jacobian determinant for the reflection case is of the form

$$J = -1 + x \theta_y - y \theta_x$$

Clearly, the Jacobian for the rotation-translation case is $J = 1$ when:

$$x_n \theta_y = y_n \theta_x$$

This is the area-preserving case resulting when the rotation angle is expressed by any function of the form

$$\theta_n = f\left(x_n^2 + y_n^2\right) = f\left(r_n^2\right)$$

or by functions of the radius $r_n = \sqrt{x_n^2 + y_n^2}$.

Another issue is to explore relations regarding the argument $z_{n+2} = z_n$ or:

$$(x_{n+2}, y_{n+2}) = (x_n, y_n)$$

By applying the rotation-translation equation above and the last relation yields:

$$x_{n+1} + x_n = a$$

Clearly, when $x_n = a/2$ then from the last relation $x_{n+1} = a/2$. Both points coincide as the relations $(x_{n+1} = a/2, y_{n+1})$ and $(x_n = a/2, y_n)$ imply that $y_{n+1} = y_n$. The more general case $z_{n+m} = z_n$ yields a relation of the form:

$$x_{n+m-1} + x_{n+m-2} + \cdots + x_n = m\frac{a}{2}$$

9.8.1   The space contraction rotation-translation case

This case is expressed by the well-known function

$$z_{n+1} = a + bz_n e^{i\theta_n}$$

where $0 < b < 1$ is the space contraction parameter. The Jacobian determinant of the space contraction rotation-translation case is:

$$J = b^2(1 + x_n\theta_y - y_n\theta_x)$$

The same limit argument

$$(x_{n+1}, y_{n+1}) = (x_n, y_n) = (x, y)$$

will give the relation

$$(x - a)^2 + y^2 = b^2(x^2 + y^2)$$

or, after some algebra:

$$\left(x - \frac{a}{1 - b^2}\right)^2 + y^2 = \left(\frac{ab}{1 - b^2}\right)^2$$

This is the equation of a circle with radius $R = \frac{ab}{1-b^2}$ centred at $(x, y) = \left(\frac{a}{1-b^2}, 0\right)$. The resulting relations for x and y are:

$$x = \frac{a(1 - b\cos\theta)}{1 + b^2 - 2b\cos\theta}$$
$$y = \frac{ab\sin\theta}{1 + b^2 - 2b\cos\theta}$$

The equation for $\cos\theta$ is:

$$\cos\theta = \frac{1}{2b}\left(1 + b^2 - \frac{a^2}{r^2}\right)$$

When the angle θ is a function of r, the last equation provides the values of θ for which $z_{n+1} = z_n$.

Questions and Exercises

1. Show that plane isometries preserve collinearity: Let A, B, C be three distinct collinear points. Then use the fact that the distances between TA, TB, TC must be the same as those between A, B, C respectively to show that TA, TB, TC must also be collinear, and in fact appear in the same order (possibly reversed).

2. Verify the properties of rotations and reflections given in (9.6).

3. What circles are left invariant by each of the different isometries?

4. Change the system (9.8) by adding a translation parameter a_1 to the second equation, while retaining the translation parameter a in the first equation. Find the axis of symmetry for this new system.

5. Change the system (9.8) by adding a translation parameter $-a$ to the second equation, while retaining the translation parameter a in the first equation. Find the axis of symmetry for this new system.

6. In the system from the previous question, use relation (9.12) to find a chaotic attractor of the Ikeda type. You will likely have to use several values for the parameters.

7. Repeat the previous question, but now using reflection equations.

8. Find the fixed points of the relation:

$$r_{n+1} = \frac{1}{(r_n - b)^2}$$

What is the smallest value of b for which there are three fixed points? How does the system behave when there is only one fixed point?

9. Draw chaotic simulations for the circular system described by:

$$r_{n+1} = \frac{1}{(r_n - b)^2}$$

10. Repeat the same analysis as in the two previous questions, but now using the relations

$$r_{n+1} = b - \frac{a}{r_n^2}$$

and

$$r_{n+1} = b - \frac{a}{r_n^4}$$

respectively. Try several values for the parameter a, and explore if there is a relation between a and b related to the chaotic behaviour of the system.

Chapter 10

Chaotic Advection

10.1 The Sink Problem

Questions related to chaotic advection go back to the nineteenth century and the development of hydrodynamics, especially the introduction of the Navier-Stokes equations (Claude Navier, 1821 and George Stokes, 1845). The vortex flow case and the related forms including vortex-lines and filaments, vortex rings, vortex pairs and vortex systems can be found in the classical book by Horace Lamb, first edited in 1879 (Lamb, 1879). However, the formulation of a theory that partially explains the vortex problem and gives results that coincide with the real life situations has only been achieved in recent years, thanks to progress in computer simulations. The introduction of terms like chaotic advection and the blinking vortex system appeared only in last decades, in order to define and analyse specific vortex flow cases.

In most cases, the formulation and solution of the problem followed the differential equations approach, mostly directed towards the solution of a Navier-Stokes boundary value problem. Few interesting cases are based on difference equations analogues. However, the formulation and analysis of vortex flow problems by means of difference equations can be very useful for several cases if a systematic study is applied. In this chapter we follow the difference equations methodology by introducing rotation-translation difference equations and a non-linear rotating angle, along with a space contraction parameter, in order to study chaotic advection problems. The interconnections between the difference and the differential equations case is also studied in specific cases.[1]

10.1.1 Central sink

Consider a circular bath with a sink in the centre at $(x, y) = (0, 0)$. The water inside the bath is rotating counterclockwise. A coloured fluid is injected in the periphery of the bath. The question we propose to address is to determine the shape of the fluid filaments if the sink is open. Geometrically the problem is that of rotation with contraction by a parameter $b < 1$. The rotation-translation model (9.10) is applied,

---

[1]The first three sections of this chapter have previously appeared in Skiadas (2007).

with the translation parameter a being zero. The equations of flow would then be:

$$
\begin{aligned}
x_{t+1} &= b(x_t \cos \phi_t - y_t \sin \phi_t) \\
y_{t+1} &= b(x_t \sin \phi_t + y_t \cos \phi_t)
\end{aligned}
\tag{10.1}
$$

The contraction in the radial direction ($r = \sqrt{x^2 + y^2}$) is found from system (10.1):

$$
r_{t+1} = b\sqrt{x_t^2 + y_t^2} = br_t
\tag{10.2}
$$

The rotation angle is assumed to follow a function of the form

$$
\phi_t = c + \frac{d}{r_t^2}.
\tag{10.3}
$$

The space contraction is given by estimating the Jacobian of the flow $J = b^2$. When $b < 1$, a particle is moving from the periphery of the bath to the sink in the centre of coordinates following spirals, as illustrated in Figure 10.1(a). The parameters are $b = 0.85$, $c = 0$, $d = 0.4$, and the initial point is $(x_0, y_0) = (1, 0)$.

When the same case is simulated for particles entering from the periphery of the rotating system at time $t = 0, 1, 2, \ldots$, Figure 10.1(b) results. The spiral forms start from the periphery and are directed towards the central sink. It is also interesting that full concentric circles appear while the spiraling flow continues. These circles have a smaller diameter, or even disappear completely, when the rotation parameter d gets smaller. The parameter b also influences the spiral. Figure 10.1(c) illustrates an advection case when $b = 0.95$, $c = 0$ and $d = 0.01$.

 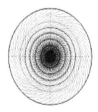 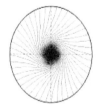

(a) Spiral particle paths (b) Spiral forms directed to the sink (c) Spiral formation toward a central sink

FIGURE 10.1: Spiral forms

10.1.2 The contraction process

From the rotation-contraction equations (10.1) and (10.2), it follows that the *radial contraction* is:

$$\Delta r_t = r_{t+1} - r_t = b r_t - r_t = -(1 - b) r_t$$

The differential equation for the contraction process is found by observing that:

$$\frac{dr}{dt} \approx \frac{\Delta r}{\Delta t} = \frac{r_{t+1} - r_t}{(t+1) - t} = -(1 - b)r$$

The resulting differential equation expressing the radial speed

$$\dot{r} = -(1 - b)r$$

is solved to give

$$r = r_0 e^{-(1-b)t}$$

where r_0 is the initial radius.

The movement is then totally determined, as the equation for the rotation angle has been given earlier. The paths form spirals towards the centre. When the movement covers a full circle, the new radius will be:

$$r = r_0 e^{-(1-b)\frac{2\pi}{\phi}}$$

10.2 Non-Central Sink

In the following case, a circular bath with a non-central sink is examined. The sink is located at $(x, y) = (a, 0)$. The equations of flow are:

$$\begin{aligned}
x_{t+1} &= b((x_t - a)\cos\phi_t - y_t\sin\phi_t) \\
y_{t+1} &= b((x_t - a)\sin\phi_t + y_t\cos\phi_t)
\end{aligned} \tag{10.4}$$

The rotation angle is assumed to follow an equation of the form

$$\phi_t = c + \frac{d}{r_t^2}$$

where $r_t = \sqrt{(x_t - a)^2 + y_t^2}$. The fixed points for this map are given by the equation

$$x^2 + y^2 = b^2\left((x - a)^2 + y^2\right)$$

or, after transformation:

$$\left(x + \frac{ab^2}{1 - b^2}\right)^2 + y^2 = \left(\frac{ab}{1 - b^2}\right)^2$$

This is the equation of a circle with radius $R = \frac{ab}{1-b^2}$ centred at $(x,y) = \left(\frac{ab^2}{1-b^2}, 0\right)$.

The flow is not symmetric. The coloured fluid starting from the outer periphery of the bath approaches the sink in few time periods, as illustrated in Figure 10.2(a). The parameters are $a = 0.15$, $b = 0.85$, $c = 0$ and $d = 0.1$. To simplify the process, it is assumed that the coloured fluid is introduced simultaneously in the periphery of the bath. Then, the circular form of the original coloured line is gradually transformed into a chaotic attractor located at the sink's centre $(x,y) = (a,0)$. The attractor is quite stable in terms of form and location.

The attractor appears even if the coloured particles are introduced into a small region of the bath, as presented in Figure 10.2(b). The coloured particles are introduced in a square region (0.1×0.1) at the right end of the bath at $(x,y) = (1,0)$. The parameters are $a = 0.15$, $b = 0.85$, $c = 0$ and $d = 0.8$. As the vortex parameter d is higher than in the previous case, the chaotic attractor appears at the 6th time step of the process. The attractor is also larger than in the previous case.

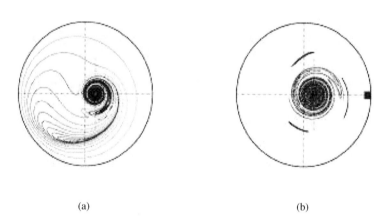

(a)   (b)

FIGURE 10.2:   Chaotic attractors in a non-central sink

10.3   Two Symmetric Sinks

10.3.1   Aref's blinking vortex system

Chaotic mixing in open flows is usually modeled by using the *blinking vortex-sink system*, invented by Aref (see Aref, 1983, 1984; Aref and Balachandar, 1986). Aref's system models the out-flow from a large bath tub with two sinks that are opened in an alternating manner, in order for chaotic mixing to take place in the course of the process. To model the velocity field caused by a sink, we assume the superposition of the potential flows of a point sink and a point vortex. If $z = x + iy$ is the complex coordinate in the plane of flow, the *complex potential* for a *sinking vortex point* is

$$w(z) = -(Q + iK) \ln |z - z_s|$$

where $z_s = (\pm a, 0)$, $2\pi Q$ is the *sink strength* and $2\pi K$ the *vortex strength*. The imaginary part of $w(z)$ is the stream function:

$$\Psi = -K \ln r - Q\phi$$

The streamlines are logarithmic spirals defined by the function:

$$\phi = -(K/Q) \ln r + \text{const}$$

The differential equations of motion in polar coordinates are

$$\dot{r} = -\frac{Q}{r}$$
$$r\dot{\phi} = \frac{K}{r} \tag{10.5}$$

and their solutions are:

$$r = \sqrt{r_0^2 - 2Qt}$$
$$\phi = \phi_0 - \frac{K}{Q} \ln \left(\frac{r_t}{r_0} \right) \tag{10.6}$$

The flow is fully characterised by the *adimensional sink strength* η and the ratio of vortex to sink strength ξ, given by:

$$\eta = \frac{QT}{a^2}, \qquad \xi = \frac{K}{Q} \tag{10.7}$$

The parameter T is the *flow period*, usually set to 1, and a is the distance of each sink from the centre of coordinates. As indicated in the literature (Károlyi and Tél, 1997; Károlyi et al., 2002), chaotic flow appears for parameter values $\eta \geq 0.5$ and $\xi = 10$. More precisely, when particles are injected into the flow in few time periods, they are attracted to a specific region (the attractor) of the flow system. In recent years, several studies have appeared investigating this phenomenon both theoretically and experimentally. The theoretical studies also include simulations using large grids (1000×1000), as well as numerical solutions of the general equations of flow. These studies suggest that the attractors are time periodic according to the time periodicity of the flow. However, if only one sink is used, a stable attractor could be present, both theoretically and experimentally, as the simulations presented above show.

We apply Aref's blinking vortex system first of all to a rotating fluid. We select a counterclockwise rotation. The symmetric sinks are located at $(x, y) = (-a, 0)$ and

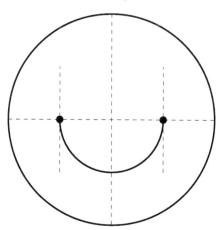

FIGURE 10.3:   The two symmetric sinks model

$(x, y) = (a, 0)$ and the time period is $T = 1$. According to this system, the flow is not stationary and there are jumps in the velocity field at each half period $T/2$. In other words, a particle located at $(-a, 0)$ appears at $(a, 0)$ in the next time period, as illustrated in Figure 10.3.

    We propose to analyse a discrete time system, like Aref's system, by using the theory of difference equations and discrete systems. It is more straightforward to consider the geometry of this system. The model we search must be a rotation-translation one with a contraction parameter $b < 1$ expressing the gradual shortening of the radius r that causes the particles to follow logarithmic spiral trajectories around the sinking vortices. In accordance with the theory just discussed, a rotation-translation model of this type is expressed by the difference equation:

$$z_{t+1} = a + b(z_t - z_s)e^{i\phi_t} \tag{10.8}$$

Equation (10.8) can be written as

$$x_{t+1} + iy_{t+1} = a + b((x_t + a) + iy_t)(\cos \phi_t + i \sin \phi_t) \tag{10.9}$$

The system of iterative difference relations for x and y is obtained by equating the real and imaginary parts of (10.9):

$$\begin{aligned} x_{t+1} &= a + b(x_t + a) \cos \phi_t - by_t \sin \phi_t \\ y_{t+1} &= \quad b(x_t + a) \sin \phi_t + by_t \cos \phi_t \end{aligned} \tag{10.10}$$

    If a particle is located at $(x_0, y_0) = (-a, 0)$, then after time $t = 1$ it will be located at $(x_1, y_1) = (a, 0)$. We next need to specify the form of the angle ϕ_t. We will use the form:

$$\phi_t = \phi_0 + \frac{\eta\xi}{r^2} = c + \frac{d}{r^2} \tag{10.11}$$

where $c = \phi_0 = \pi$ is chosen to account for the half-cycle rotation, $d = \eta\xi$ is the vortex strength, and $r = r_t$ is the distance from the vortex $(-a, 0)$, namely:

$$r = \sqrt{(x + a)^2 + y^2}$$

For the experiments presented in the literature, $\eta = 0.5$, $\xi = 10$ and thus $d = 5$. However, the chaotic region is wider, as illustrated in the following figures.

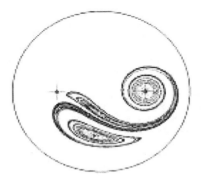

FIGURE 10.4:   Chaotic attractor in the two-sink problem

The chaotic attractor in Figure 10.4 illustrates the two-sink case for parameter values $a = 1$, $b = 0.8$, $c = \pi$ and $d = 3$. There are two main vortex forms counterbalancing each other. The first form is located at the right-hand-side sink at $(x, y) = (a, 0)$. The second vortex form is centred at $(x, y) = (a + 2ab\cos\phi, 2ab\sin\phi)$, where $\phi = d/(4a^2)$. The two main vortex forms can be separated when the parameter d expressing the vortex strength is relatively small. Such a case is presented in Figure 10.5. The parameter d here is set to 1, while the other parameters remain the same as in the previous example. The attractor is now completely separated into two chaotic vortex forms (attractors).

Another possibility is to give high values to the parameter d expressing the vortex strength. The value $d = 2\pi$ leads to a more complicated vortex form, as presented in Figure 10.6. There are three equilibrium points, for times $t = 1, 2, 3$. The first of these points is the centre of the right-hand-side sink.

One can also find the form of the vortices in the case of three sinks located in an equilateral triangle. The simulation of this situation is achieved by selecting a value $c = \frac{2\pi}{3}$ for the sink strength. The other parameters are $b = 0.95$, $d = \pi$ and $a = 5$. Figure 10.7(a) illustrates this case.

A form with four vortices is presented in Figure 10.7(b). This is achieved by assuming a special value, $c = \frac{2\pi}{4}$, for the parameter c, thus dividing the total circle in four sectors. The positions of the four sinks are located on the corners of a square.

FIGURE 10.5:   Two distinct vortex forms ($d = 1$)

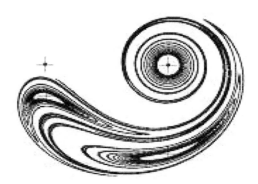

FIGURE 10.6:   The chaotic attractor with strong vortex strength parameter
$d = 2\pi$

The parameters selected for the simulation are $b = 0.98$, $d = \pi$ and $a = 5$. Similarly, a fifth order vortex form results when $c = \frac{2\pi}{5}$ and the other parameters remain unchanged (Figure 10.7(c)).

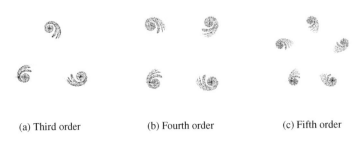

(a) Third order   (b) Fourth order   (c) Fifth order

FIGURE 10.7:   Vortex forms

10.4   Chaotic Forms without Space Contraction

Chaotic forms similar to those produced earlier in this chapter, but without any space contraction, *i.e.* when $b = 1$, resulting in a Jacobian of $J = 1$, are of particular interest. As there is no radial or space contraction, the flow could be chaotic, but the resulting chaotic attractors are not classified as main vortex forms.

Figure 10.8(a) illustrates the case where $a = 1$, $b = 1$, $c = \pi$ and $d = 1$. The chaotic form is symmetric. The axis $y = 0$ is the symmetry axis. A division of the main chaotic form is starting. As the vortex strength parameter d is decreasing ($d = 0.4$ for Figure 10.8(b) and $d = 0.3$ for Figure 10.8(c)) two distinct symmetric chaotic forms arise. Figure 10.8(d) gives an enlargement of the right-hand-side chaotic attractor produced when $d = 0.3$. Interesting chaotic details are present in this attractor.

10.5   Other Chaotic Forms

In the following simulation, we use the classical rotation-translation model with a rotation angle expressed by an inverse function of the square of the radius r:

$$\theta_n = c + \frac{d}{x_n^2 + y_n^2}$$

We examine the effect of varying the rotation parameter c while keeping the other parameters fixed. The space contracting parameter b takes a value close to 1 ($b = 0.99$), which could express low speeds directed towards the sink. When $a = 10$, $d = 10$ and $c = 1$, six rotating vortex-like spiral objects appear (Figure 10.9(a)).

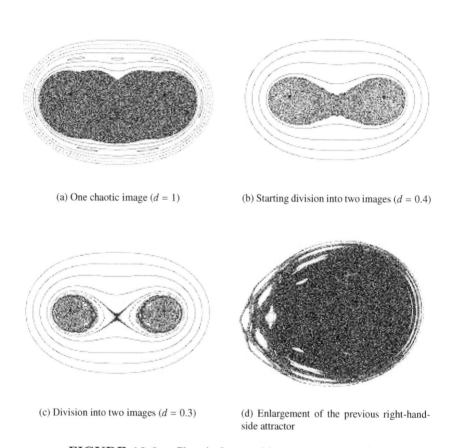

(a) One chaotic image ($d = 1$)   (b) Starting division into two images ($d = 0.4$)

(c) Division into two images ($d = 0.3$)   (d) Enlargement of the previous right-hand-
                                             side attractor

FIGURE 10.8:   Chaotic forms without space contraction

When $c = 2$, with the other parameters unchanged, three objects appear, as illustrated in Figure 10.9(b). These chaotic forms illustrate two-armed spiral galaxies located at

$$(x, y) = (0, 0),\ (x, y) = (a, 0) \text{ and } (x, y) = (a + ba\cos(\theta_a),\, ba\sin(\theta_a))$$

where:

$$\theta_a = c + \frac{d}{a^2}$$

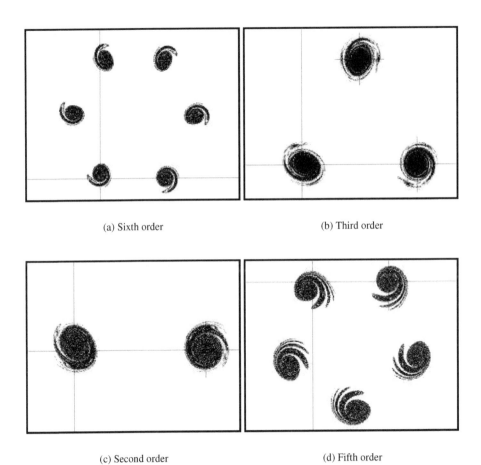

(a) Sixth order   (b) Third order

(c) Second order   (d) Fifth order

FIGURE 10.9: Higher order vortices

For $c = 3$, the chaotic form of the attractor is two one-armed spirals, illustrated in Figure 10.9(c). The chaotic attractors are located at $(x, y) = (0, 0)$ and $(x, y) = (a, 0)$. Five chaotic attractors appear when $c = 5$ (Figure 10.9(d)).

When c takes smaller values, for instance $c = 0.5$, multi-armed spirals are present, as seen in Figure 10.10(a). The original rotating disk has radius $R = 1$. When $R = 2$, Figure 10.10(b) appears. The number of spirals remains the same, but more details are present in Figure 10.10(c). The original rotating disk has radius $R = 3$.

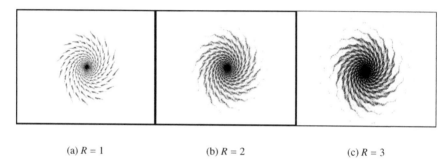

(a) $R = 1$ (b) $R = 2$ (c) $R = 3$

FIGURE 10.10:   Spiralling towards the sink

10.6   Complex Sinusoidal Rotation Angle

Very interesting chaotic attractors arise by assuming that the rotation angle of the rotation-translation model is a complex sinusoidal function. For the first application, the following function for the rotation angle is considered:

$$\theta_n = \frac{1}{\sin\left(c - \frac{d}{1 + x_n^2 + y_n^2}\right)}$$

The parameters are set to $a = 1$, $b = 0.83$, $c = 0.4$ and $d = 6$. Figure 10.11 illustrates the resulting chaotic attractor. The image shows a circular ring along with an internal vortex form and an external part consisting of three main vortex forms and a smaller fourth one.

A change in the function for the rotation angle results in the interesting chaotic rotation image presented in Figure 10.12(a). The parameters remain the same as in

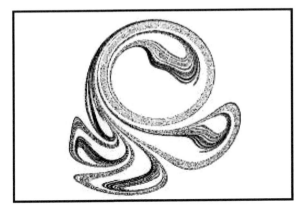

FIGURE 10.11: A sinusoidal rotation angle chaotic attractor ($c = 0.4, d = 6$)

the preceding example, while the rotation angle function now is:

$$\theta_n = \sin\left(c - \frac{d}{\sqrt{x_n^2 + y_n^2}}\right)^{-1} = \sin\left(c - \frac{d}{r_n}\right)^{-1}$$

The resulting image has a circular ring with two main vortex forms, and a smaller one in the outside region, as well as vortex forms inside, interconnected with a central rotating part.

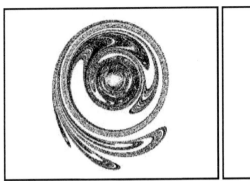

(a) $\theta_n = 1/\sin\left(c - d/r_n\right)$ (b) $d = 16$

FIGURE 10.12: Sinusoidal rotation angle chaotic attractors

A relatively simpler function for the rotation angle, with only one parameter, is:

$$\theta_n = \sin\left(\frac{d}{1 + x_n^2 + y_n^2}\right)^{-1}$$

However, the resulting chaotic image in Figure 10.12(b) is also quite complicated. As before, there is a circular ring and three connected vortex forms outside of the ring, while several vortex forms rotate clockwise and counterclockwise inside the ring. The parameters are $a = 1.2$, $b = 0.85$ and $d = 16$.

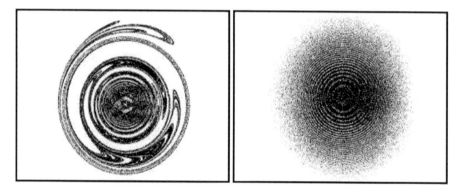

(a) $\theta = 1/\sin\left(dr^2\right)$, $b = 0.7$ (b) $\theta = 1/\sin\left(dr^2\right)$, $a = 0.85$, $b = 0.8$

FIGURE 10.13:   Sinusoidal rotation angle chaotic attractors

A completely different image results by introducing another function for the rotation angle:

$$\theta_n = \frac{1}{\sin(d\,(x_n^2 + y_n^2))} = \frac{1}{\sin\left(dr^2\right)}$$

The same rotation-translation case with parameters $a = 1.2$, $b = 0.85$ and $d = 16$ provides also a circular ring image (Figure 10.13(a)) with surrounding vortex forms, but in this case the outside ring vortex forms are small, whereas the inside part is larger, including a central circular bulge and vortex forms rotating clockwise and counterclockwise.

The standard rotation-translation model is also applied with parameters $a = 1.1$, $b = 0.8$ and $d = 16$ (Figure 10.13(b)). A large number of concentric rings appear.

10.7   A Special Rotation-Translation Model

A special rotation-translation model is given by the following set of relations

$$
\begin{bmatrix} x_{n+1} \\ y_{n+1} \end{bmatrix} = \mathbf{rot}\,\theta_n \begin{bmatrix} b\left(x_n^2 - y_n^2\right) \\ 2bx_ny_n \end{bmatrix} + \begin{bmatrix} a \\ 0 \end{bmatrix}
$$

$$
= \begin{bmatrix} a + b\left(x_n^2 - y_n^2\right)\cos\theta_n - 2bx_ny_n \sin\theta_n \\ b\left(x_n^2 - y_n^2\right)\sin\theta_n + 2bx_ny_n \cos\theta_n \end{bmatrix}
$$

where the rotation angle is given by the relation

$$
\theta_n = \frac{d}{r_n}
$$

This map deserves special attention, as the Jacobian is $J = 4b^2r^2$. When $r = 1/(2b)$, the map is area preserving, whereas when $r < 1/(2b)$ a space contraction takes place. For $r > 1/(2b)$ area expansion takes place. The parameter values selected for the application are $a = 0.8$ for the translation parameter, $b = 0.6$ for the area contraction parameter and $d = 5$ for the rotation parameter. Figure 10.14(a) illustrates the simulation results. There is a rotation centre and vortex-like forms.

10.8   Other Rotation-Translation Models

10.8.1   Elliptic rotation-translation

The following system of two difference equations provides elliptic forms when the ellipticity parameter h is not 1.

$$
x_{n+1} = a + b\left(x_n \cos\theta_n - hy_n \sin\theta_n \right)
$$
$$
y_{n+1} = b\left(\frac{1}{h}x_n \sin\theta_n + y_n \cos\theta_n \right)
$$
(10.12)

This system results is an analogue of the classical rotation-translation system, where the the rotation takes place on ellipses where the ratio of the one side to the other is h. They arise from the rotation-translation equations, after a rescaling on the x direction by h. To see this, denote by $(\bar{x}_n, \bar{y}_n)$ the new coordinates. Then $\bar{x}_n = hx_n$, and $\bar{y}_n = y_n$, hence we have:

$$
\bar{x}_{n+1} = hx_{n+1}
$$
$$
= hx_n \cos(\theta) - hy_n \sin(\theta)
$$
$$
= \bar{x}_n \cos(\theta) - h\bar{y}_n \sin(\theta)
$$

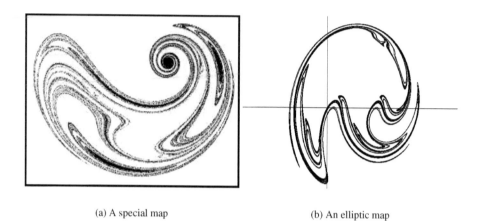

(a) A special map   (b) An elliptic map

FIGURE 10.14:   Chaotic images of rotation-translation maps

and

$$\bar{y}_{n+1} = y_{n+1}$$

$$= \quad x_n \sin(\theta) + y_n \sin(\theta)$$

$$= \frac{1}{h}\bar{x}_n \cos(\theta) - \bar{y}_n \sin(\theta)$$

Adding the translation and contraction parameters results in (10.12). This case is illustrated in Figure 10.14(b), where the rotation angle has the form $\theta = dr^2$, and the parameters are $a = 0.5$, $b = 0.9$, $d = 3$ and $h = 0.8$. The elliptic rotation equations give a somewhat deformed image, compared to the previous circular type cases presented in this chapter.

10.8.2   Rotation-translation with special rotation angle

This case is modelled by using the classical rotation-translation model with space contraction, but now the rotation angle obeys the function

$$\theta_n = c + \frac{d}{r_n^2 - e}$$

where, for our example, $e = 4.5$ and the other parameters are $a = 3, b = 0.9, c = 1.52$ and $d = 1.014$. A space contraction parameter e is inserted in the function expressing the rotation angle. The resulting image is illustrated in Figure 10.15(a). This image, and the ones following it, are of particular interest when studying chaotic advection cases.

The main part of the image is a circular ring centred at $(x, y) = (a, 0)$. Three vortex forms appear outside the circular ring and two smaller vortex forms appear inside the ring. All the vortex forms originate from the periphery of the circular ring. Smaller vortex forms appear inside every main vortex form. In the same figure, the two axes of coordinates and the part of the circle expressing the space curvature due to the space contraction parameter b and the translation parameter a are illustrated. This circle is of radius $R = \frac{ab}{1-b^2}$ centred at $(x, y) = \left(\frac{a}{1-b^2}, 0\right)$.

As the parameter c changes, the resulting image shows the same circular ring located at the same centre and having similar magnitude, but changes appear in the vortex forms. Figure 10.15(b) illustrates the case where $c = 4.3$ and $d = 1$, whereas the other parameters remain unchanged. The two vortex forms inside the circular ring take up more space, while the outer vortex forms create a main vortex image with vortex forms inside.

Figure 10.15(c) has the same parameters as the previous case, except for $c = 8.5$. There are now two outer vortex forms. However the image is similar to that of Figure 10.15(a). Figure 10.15(d) is similar, but now with $c = 17$.

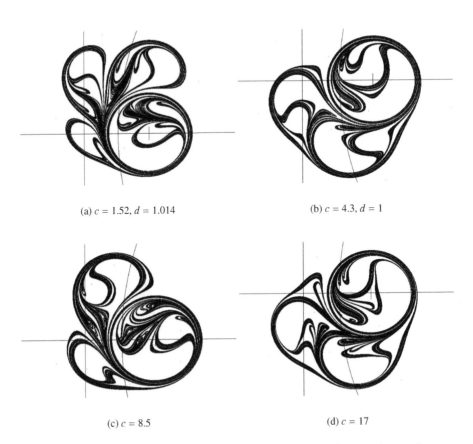

(a) $c = 1.52, d = 1.014$

(b) $c = 4.3, d = 1$

(c) $c = 8.5$

(d) $c = 17$

FIGURE 10.15: Chaotic images with area contraction rotation angles

Questions and Exercises

1. Find a symmetry axis for the system (10.10).

2. Consider the system (10.10). Add a translation parameter "k" at the second equation, and find a symmetry axis for this new system. Use the rotation angle $\phi_t = c + d/r_t^2$, and draw chaotic images when $c = \pi/2$ and d is varying.

3. Chaotic advection is closely related to *chaotic mixing*. Discuss the mixing of fluids when the chaotic region is reached. What is the disadvantage with the presence of chaos in mixing? How can we avoid chaotic phenomena in fluid mixing?

4. Change the rotation equations (10.1) to their reflection analogues, and find the resulting chaotic and non-chaotic forms for various values of the parameters b, c and d.

5. Change the rotation equations (10.10) to their reflection analogues, and find the resulting chaotic and non-chaotic forms for various values of the parameters b, c and d.

Chapter 11

Chaos in Galaxies and Related Simulations

11.1 Introduction

A large variety of chaotic forms appears in stellar observations. These forms exhibit astonishingly large varieties of shapes. At first glance, there is not a simple method to handle and organise the entire material of stellar observations based on a few rules. The related cosmology theories for star and galaxy creation and formation are of considerable interest. Related laws help us describe the evolution of stellar shape over time.

Already from the earliest observations, it was evident that chaos was the main characteristic of the shape and form of many stellar formations, including galaxies and chaotic nebulae. A great number of stellar formations develop under high temperatures, the so-called 'fire' that the ancient Greek philosopher, Heracleitos of Ephesus (Heraclitus), had emphasised as the basis of genesis. All these "fire" formations are mainly chaotic forms.

A fundamental problem in non-linear dynamics is that of exploring how the properties of orbits change and evolve as one or more parameters of a dynamical system change, in order that the system exhibit chaotic behaviour. Recent advances in non-linear physics have resulted from such an approach. The transition to chaos and the routes to chaos are extensively explored in the case of one-dimensional maps and, mainly, for the logistic map. Of considerable interest however are also the chaotic dynamics of two-dimensional maps, such as the cat-map and the Hénon map.

Another very important transition, related to physics, is one in which the system is in a chaotic state and, as parameters change, the system transitions from one type of chaos to another. Of particular interest is the case where the chaotic attractors that correspond to a system change as a parameter gradually takes higher or lower values. Many stellar and galactic systems may fit to chaotic attractor phenomena. Chaotic attractors in two-dimensional and three-dimensional space are quite stable objects and can be useful in exploring and simulating the forms of galaxies, clusters of galaxies, nebulae, black holes and other stellar objects. There are several approaches regarding the handling of chaotic situations occurring under the influence of a central force as in the case of stellar systems. An approach is to consider a system of masses moving in different orbits under gravity. Early studies of the subject have been undertaken by Contopoulos and Bozis (1964); Lynden-Bell (1969); Toomre (1963, 1964); Hunter and Toomre (1969); Contopoulos et al. (1973). A detailed approach of chaotic dynamics in astronomy can be found in Contopoulos (2002). Ostriker and Peebles (1973) explore the stability of flattened galaxies while Raha et al. (1991) study the dynamical instability of bars in disk galaxies. The box and peanut shapes generated by stellar bars are discussed in Combes et al. (1990). Extensive work on how one can separate chaotic from ordered domains is found in the works of Contopoulos and Voglis (Patsis et al., 1997). Objects of very complicated morphology are found and simulated in X-ray clusters (Navarro et al., 1995). The discovery of new ultra-luminous IRAS galaxies give rise to new studies on how galactic systems interact.

A number of works regarding chaos and chaotic attractors and the interaction with chaotic dynamics in astronomy are of interest. Most notable among them are the work of White et al. (1998) on the anomalous transport near threshold of the standard map and the approach by Arrowsmith and Place (1990); Acheson (1997) in their books on dynamical systems and calculus and chaos. Hénon and Heiles (1964) developed a Hamiltonian system which developed into a prototype for computer experiments of chaotic dynamics in astronomy. Of considerable interest is also the area-preserving quadratic map developed by Hénon (1969). Period doubling bifurcations of this map appear in recent studies (Murakami et al., 2002).

It is very complicated to express chaos analytically, namely so that every stage is explained and defined in a deterministic way. However, the forms resulting from billions of chaotic intermediate steps are frequently of considerable 'attracting' shape. These attractors are in many cases reproduced in the laboratory by using very simple rules.

Following the studies on chaos in the last decades, it is evident very few main characteristics or parameters of the chaotic system play a critical role in the formation of attractors. In stellar systems the main characteristics are: 1) gravity, the attracting force between bodies (masses), and 2) the first Big-Bang. Geometrically, stellar systems move, grow or decline by: a) rotating and b) translating (following linear movements). According to the established theories, expansion of the universe (translation) follows the original Big-Bang, whereas gravity, the attracting force, causes translation in the opposite direction. Attracting forces in different directions cause rotations and translations in stellar formations.

In chaotic dynamics, the main effort is usually centred on the formulation of relatively simple models with only a few chaotic parameters. If the main characteristics of the phenomenon are included in the final model, the chaotic character of many phenomena in nature could be well expressed. Chaotic formations in galaxies could appear when rotations, reflections and translations (axial movements) are present along with the influence of gravitational forces that play an important role. In this chapter, simple iterative models are presented. They are the simplest possible models, including rotation, reflection, translation and the influence of gravitational forces on the angle of rotation. These models present interesting aspects that are studied analytically and are presented in several graphs following the simulation. Several spiral forms of various formations can be modelled by changing the parameters of the models. The study of galactic forms by using differential equations and Hamiltonians is presented in Chapter 12.

11.2   Chaos in the Solar System

Complicated and even chaotic orbits and paths of the solar system are studied and simulated in related chapters of this book. The bodies in the solar system mainly follow cyclic or elliptic orbits around a centre of mass. In this case, the simulation can follow a purely geometric approach. To simplify the problem, the centre of mass is located in the centre of the large mass. The attracting force is, of course, gravity, and it is presented by an inverse square law for the distance. The non-dimensional equations of motion for a body with small mass m rotating around a large body with mass M, which we consider to stay in equilibrium position, are, in Cartesian coordinates:

$$\ddot{x} = -\frac{x}{(x^2 + y^2)^{3/2}}$$
$$\ddot{y} = -\frac{y}{(x^2 + y^2)^{3/2}}$$

(11.1)

The initial conditions are $x = 1, y = 0, \dot{x} = 0$ and $\dot{y} = v$. The escape velocity is $v = \sqrt{2}$, and the circular orbit is obtained when $v = 1$. The circular orbit is illustrated in Figure 11.1(a). In the same figure an elliptic orbit is presented, where the initial velocity is $1 < v = 1.2 < \sqrt{2}$. A hyperbolic orbit is also presented. The initial velocity is higher than the escape limit ($v = 1.45 > \sqrt{2}$) and the particle flies off to infinity. Figure 11.1(b) illustrates a large number of elliptic paths for the same value of the initial velocity as before. The paths remain in the same plane, but deviate from the original position, forming the shape presented in the figure.

The motions of two attracting point masses are relatively simple. The equations of motion for the point mass m_1 in Cartesian coordinates are:

$$\ddot{x}_1 = -\frac{(x_1 - x_2)m_2}{((x_1 - x_2)^2 + (y_1 - y_2)^2)^{3/2}}$$
$$\ddot{y}_1 = -\frac{(y_1 - y_2)m_2}{((x_1 - x_2)^2 + (y_1 - y_2)^2)^{3/2}}$$

(11.2)

The case for the point mass m_2 is analogous.

Without loss of generality, the gravity constant is $G = 1$ and the total mass of the rotating system is $M = m_1 + m_2$. If we give the two masses initial positions (x_1, y_1) and (x_2, y_2), and initial velocities $\dot{x}_1, \dot{y}_1, \dot{x}_2$ and $\dot{y}_2$, the masses revolve around each other while drifting together, as a pair, in space. Their centre of mass drifts at a constant velocity. Relative to an observer moving with the centre of mass, their orbits are ellipses and, in special cases, circles.

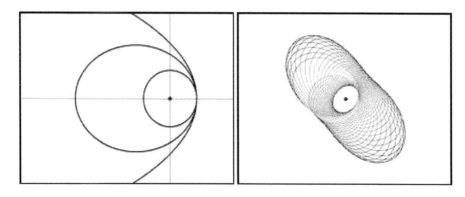

(a) Circular and elliptic orbits   (b) Elliptic paths

FIGURE 11.1:   Elliptic orbits

The motion of two attracting points in space is presented in Figure 11.2(a). The heavy line represents the movement of the centre of mass. The masses revolve around and drift together. The points have masses $m_1 = 0.4$, $m_2 = 0.6$, and initial velocities $\dot{x}_1 = -0.15$, $\dot{x}_2 = 0$, $\dot{y}_1 = -0.45$ and $\dot{y}_2 = 0.25$. The same motion, viewed relative to the centre of mass, is illustrated in Figure 11.2(b). Each point-mass follows an elliptic path.

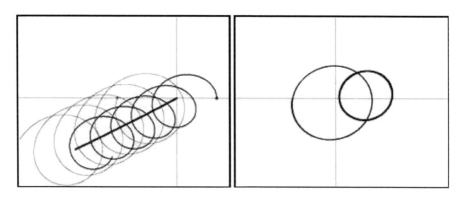

(a) Attracting bodies in space   (b) Relative movement

FIGURE 11.2:   Attracting bodies in space and relative movement

The three-body problem reduced in the plane still gives very complicated and chaotic orbits. To simplify the problem, we assume that the third body, a satellite, is of negligible mass related to the other two bodies. The other two, the planets (primaries), travel about the centre of the combined mass in circular orbits in the same plane. The satellite rotates in the same plane. The total mass of the planets is $m_1 + m_2 = 1$. The equations of motion for the satellite are

$$\ddot{x} = -\frac{m_1(x - x_1)}{r_1^{3/2}} - \frac{m_2(x - x_2)}{r_2^{3/2}}$$

$$\ddot{y} = -\frac{m_1(y - y_1)}{r_1^{3/2}} - \frac{m_2(y - y_2)}{r_2^{3/2}}$$

(11.3)

where (x, y) is the position of the satellite, r_1, r_2 are the distances of the satellite from the two planets m_1, m_2 respectively, and (x_1, y_1) and (x_2, y_2) are the positions of the planets at the same time t.

In the following example, the primaries have masses $m_1 = 0.9$ and $m_2 = 0.1$, and the position and initial velocity of the satellite are $(-1.033, 0)$ and $(0, 0.35)$ respectively. The complicated and chaotic paths that arise are illustrated in the following figures. In Figure 11.3(a), the movement is viewed from the outside of the system, from "space." It shows that the satellite can move between the two revolving primaries. Figure 11.3(b) illustrates the orbit of the satellite viewed in a coordinate system that rotates with the revolution of the planets (rotating frame), with the x-axis always passing through the planets, according to the rotation formula:

$$x_{new} = x\cos t + y\sin t$$

$$y_{new} = -x\sin t + y\cos t$$

(11.4)

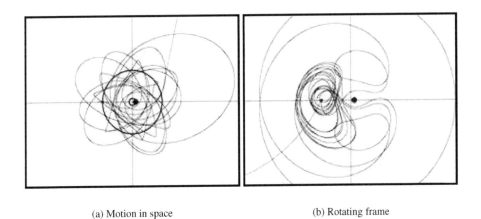

(a) Motion in space   (b) Rotating frame

FIGURE 11.3: Motions in space

By using difference equations to model solar systems, it is possible to simulate chaotic patterns like, for instance, the rings of Saturn and of other planets. The simplest approach is to use an equation such as the logistic, and to assume that the system rotates following elliptic orbits. Two realisations appear in the next figures. The parameter of the logistic model is $b = 3.6$ for Figure 11.4(a) and $b = 3.68$ for Figure 11.4(b). The rotation formula is given by:

$$x_t = 2r_t \cos t$$
$$y_t = 2r_t \sin t$$
(11.5)

where $r_{t+1} = br_t(1 - r_t)$.

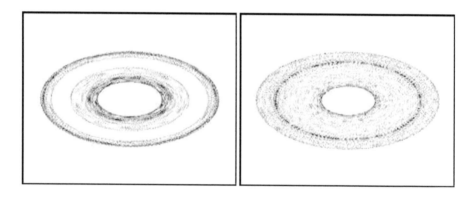

(a) $b = 3.6$ (b) $b = 3.68$

FIGURE 11.4:   Ring systems

11.3   Galaxy Models and Modelling

Studies on chaotic dynamics in galaxies and in other complicated stellar objects indicate that order and chaos coexist in these structures. It is also quite difficult to simulate complicated stellar objects, especially in the limits where chaos is present. Experience on chaotic simulations during the last decades indicates that the simpler the chaotic model, the more successful it is. Remarkable solutions of the flow equations were provided by Lorenz (1963). He transformed a high-dimensional system into a three-dimensional one, retaining the chaotic character of the system. Rössler (1976d) proposed the simplest model of three-dimensional flow. Since, even nowadays, the subject of chaos in astronomy has many unexplored aspects, a convenient way of dealing with the problem is to explore the qualitative and quantitative aspects by considering simple chaotic models, in which only the main characteristics of the system under consideration are retained. Such characteristics in galaxies and galactic formations are of course: 1) rotation-reflection, 2) translation, and 3) area contraction due to gravitation (expressed by a parameter $b < 1$). Rotation is connected with the theory of a central force by assuming that the rotation angle obeys a function of the distance r from the origin. Translation is assumed to take place when a strong gravitation source is located in a large distance from the origin or, even, when gravity waves are present. To simplify the problem, only the two-dimensional approach in the (x, y) plane is presented here.

A convenient formulation of the iterative map expressing rotation-reflection and translation is given by the following difference equation:

$$f_{n+1} = a + b f_n e^{i\theta}$$

where:

$$f_n = x_n + iy_n \qquad \text{(for rotation)}$$
$$f_n = x_n - iy_n \qquad \text{(for reflection)}$$

For rotation, the determinant of this map is $\det J = b^2 = \lambda_1 \lambda_2$. This is an area contracting map if the Jacobian J is less than 1. The eigenvalues of the Jacobian are λ_1 and λ_2. For reflection the Jacobian gives: $\det J = -b^2$. The translation parameter is a in both cases.

In the cases discussed above, rotation and reflection are followed by a translation along the x-axis. As discussed earlier, this has the advantage, over a general translation, of reducing the number of parameters by one, without affecting the results in the majority of the cases studied.

The angle θ is assumed to be a function of the distance r from the origin. In the cases studied here, $r = \sqrt{x^2 + y^2}$, the Euclidean metric. The general forms studied can be classified as follows: $\theta = g(r^m)$, where $m = 1, 2, -1, -2, -3/2$, are the most convenient values for the exponent. By expanding the function $g(r^m)$ in a Taylor series and retaining the first terms to the right the following forms for the angle θ for reflection and rotation cases arise:

$$\theta_n = c r_n$$
$$\theta_n = c r_n^2$$
$$\theta_n = c + \frac{d}{r_n}$$
$$\theta_n = c + \frac{d}{r_n^2}$$
$$\theta_n = c + \frac{d}{r_n^{2/3}}$$

Both rotation and reflection generate chaotic attractors for appropriate values of the parameters a, b, c, and d.

A discussed in Chapter 8, a very interesting property of the map is the existence of a disk F of radius $\frac{ab}{1-b}$, centred at $(a, 0)$ in the (x, y) plane, the points of which remain inside of it under the map. Further, the fixed points obey the equations

$$\left(x_n - \frac{a}{1-b^2}\right)^2 + y_n^2 = \frac{a^2 b^2}{(1-b^2)^2}$$

and, for rotation:

$$\cos \theta_n = \frac{1}{2b}\left(1 + b^2 - \frac{a^2}{r_n^2}\right)$$

So the fixed points lie on a circle K with radius $R = \frac{ab}{1-b^2}$, centred at $(\frac{a}{1-b^2}, 0)$ in the (x, y) plane.

For the first application we choose the following function for the rotation angle:

$$\theta_n = c + \frac{d}{r^2}$$

The simulation results are illustrated in Figures 11.5(a) through 11.5(f). In these figures, the parameters of the model are $b = 0.8$, $c = 3$ and $d = 6$. This is an example where an area contraction takes place, equal to $b^2 = 0.64$. The other parameter, a, varies, starting from very low values, ($a = 0.01$ in Figure 11.5(a)) and taking the values $a = 0.45$, $a = 1.6$, $a = 2.4$, $a = 4$ and $a = 5$ in the subsequent figures.

In Figure 11.5(a), the parameter a seems to have very little effect in the development of the phenomenon. The point masses are distributed around the centre of coordinates and their distribution follows an inverse law. In the second case (Figure 11.5(b)), a rotating form appears. This chaotic form is restricted inside the two circles F and K that were discussed earlier. Higher values of a, as in Figure 11.5(d), produce chaotic forms in which a centre of revolution appears in $(a, 0)$, but further a quite large rotating form appears which, when a takes higher values (Figure 11.5(c) and Figure 11.5(e)), leads to the formation of two distinct attractors. In the case of Figure 11.5(f), these attractors are finally separated in two forms. The centres of the two attractors are located at $(a, 0)$ for the attractor on the right side of Figure 11.5(f), and at

$$\left(a + a\cos\left(c + \frac{d}{a^2}\right), a\sin\left(c + \frac{d}{a^2}\right)\right)$$

for the attractor on the left side. An interesting property of the above simulation is that the original single simple rotating form splits into two forms, that show the same direction of rotation and are located at places depending on the displacement a.

More complicated chaotic forms arise for appropriate values of the parameters a, b, c and d. When $a = c = d = 4$ and $b = 0.9$, three attractors appear, one of which is located at $(a, 0)$. This is illustrated in Figure 11.6(a).

Another example appears in Figure 11.6(b), presenting a cluster of chaotic attractors. The first attractor, located at $(a, 0)$, is at the top of the figure. The second attractor is the one on the right side of the figure, while the other attractors continue in a clockwise direction until the final point, located at the circle K. The parameters are: $a = c = d = 5.25$ and $b = 0.8$.

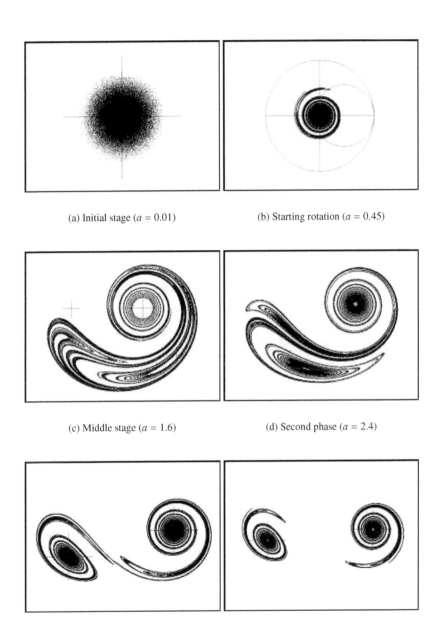

(a) Initial stage ($a = 0.01$)   (b) Starting rotation ($a = 0.45$)

(c) Middle stage ($a = 1.6$)   (d) Second phase ($a = 2.4$)

(e) Two interconnected images ($a = 4$)   (f) Two separate images ($a = 5$)

FIGURE 11.5:   Chaotic forms of a simple rotation-translation model

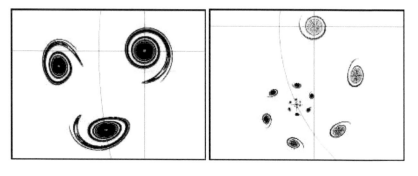

(a) Three chaotic images   (b) Multiple chaotic images

FIGURE 11.6:   Chaotic forms of a simple rotation-translation model

Rotating forms can also be formed by using rotation equations with a linear expression for the rotation angle:

$$\theta_n = c + dr_n$$

The translation parameter is a, and an area contracting parameter $b < 1$ is essential to keep the system into a limit included within a circle F of radius $\frac{ab}{1-b}$ centred at $(a, 0)$. A number of rotating point masses are equally distributed within the circle F, and the question is how these points will be distributed after time t corresponding to n iterations. The results are summarised in Figures 11.7(a) to 11.7(d). A two-armed spiral structure appears, located at $\left(\frac{a}{2}, \lim (y_n = y_{n+1})\right)$, and four satellites are located at $(\lim(x_n), \lim(y_n))$. Two of the satellites have their location centre at the circumference of the circle K of radius $\frac{ab}{1-b^2}$. The centre of this circle is at $\left(\frac{a}{1-b^2}, 0\right)$. In Figures 11.7(a), 11.7(b) and 11.7(c) the parameters are $a = 4.25$, $b = 0.9$, $c = 1.5$ and $d = 0.3$. In Figure 11.7(d) the parameters are $a = 2.3$, $b = 0.9$, $c = 2.03$ and $d = 0.3$. In these cases there are attracting points and no other attractors. The point masses are trapped in trajectories and are led to the attracting points as when a black hole is present. The attracting points are solutions of the equations that satisfy the special conditions set for the general rotation equation above.

Chaotic attractors and attracting points appear in an example of rotation (Figure 11.8(a)) where the rotation angle follows an inverse law regarding the distance r from the origin:

$$\theta_n = c + \frac{d}{r_n}$$

In the first case, the translation parameter is $a = 0.6$ and the other parameters are $b = 0.9$, $c = 3$ and $d = 6$. The chaotic attractor is located at $(a, 0)$ and the attracting point is located in the periphery of the circle K. The rotating points inside the circle F are trapped in the attractor region, or they disappear into the attracting point. The points that oscillate inside the attracting region escape from this region only if they

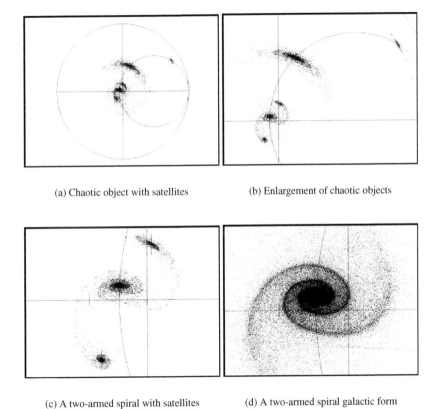

(a) Chaotic object with satellites   (b) Enlargement of chaotic objects

(c) A two-armed spiral with satellites   (d) A two-armed spiral galactic form

FIGURE 11.7: Chaotic forms of a simple rotation-translation model

reach the periphery of the attractor, and then they follow the trajectory that leads to the attracting point.

The same behaviour is seen in the next example (Figure 11.8(b)), where the parameters are $a = 5$ and $c = 2$, while the other parameters remain unchanged. Now two attractors appear, and each attractor is connected to the attracting point with trajectories. In Figure 11.8(b), the attracting points are the points where $x_{n+2} = x_n$ and $y_{n+2} = y_n$, while, in Figure 11.8(a), the attracting point is the point where $x_{n+1} = x_n$ and $y_{n+1} = y_n$.

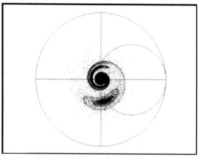
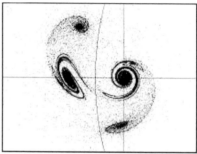

(a) Attractor and attracting point

(b) Two attractors and the accompanying attracting points

FIGURE 11.8:   Attractors and attracting points in rotation-translation model

11.3.1   A special rotation-translation image

This example comes from Contopoulos and Bozis (1964). The situation examined was that of two galaxies approaching each other. Here, we apply a rotation-translation model with space contraction. The rotation angle and the other parameters, except for the space contraction parameter, are the same as in Contopoulos and Bozis (1964). The rotation angle has the form:

$$\theta = -d\left(\frac{1}{6}\ln\frac{r^2 - cr + c^2}{(r + c)^2} - \frac{1}{3}\tan\frac{2r - c}{c\sqrt{3}}\right)$$

The parameters for the system are $a = 4$ for translation, and $c = \frac{1}{0.012^{1/3}}, d = \frac{0.0055}{0.012^{2/3}}$ and $b = 0.9$ for space contraction. The resulting shape is illustrated in Figure 11.9. A number of particles are introduced into a small circle near the centre of coordinates. The particles follow characteristic trajectories over time. The image is quite complicated. There are several attracting points, as well as places with high and low density.

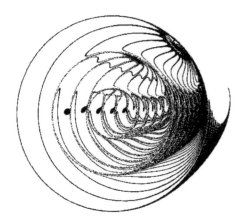

FIGURE 11.9:   A rotation image following a Contopoulos-Bozis paper
(approaching galaxies)

11.4   Rotation-Reflection

It is interesting to explore the relationships between chaotic attractors in the presence of rotation and reflection respectively, for the same set of parameters. One would expect that the attractors for the reflection case would be a mirror image of the attractors for the rotation case. This can be seen in some particular cases, for instance in the following, where the rotation or reflection angle is of the form $\theta_n = dr^2$. The parameters in both cases (Figure 11.10(a) illustrates rotation and Figure 11.10(b) illustrates reflection) are, $a = 2$, $b = 0.5$ and $d = 0.65$. The chaotic forms of the attractors verify the assumption that there is a mirror-image relationship between the two.

The chaotic geometric modelling approach presented in this section could be useful in galaxy simulations. Chaotic attractors and other forms that appear in real space (x, y) simulations indicate that elements of any form can be trapped inside these attractors. When the space contraction parameter $b < 1$ three cases appear, according to the magnitude of the rotation angle θ:

1. the particles lead to attracting points where they disappear as if in a black-hole,

2. in the presence of a chaotic attractor, the particles will stay or oscillate inside this attractor, and

3. the particles follow trajectories leading away from the rotating system if the translation parameter is relatively high.

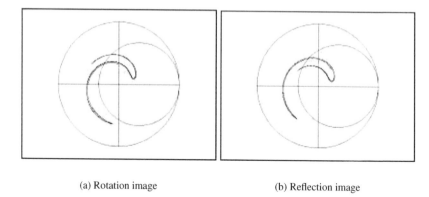

(a) Rotation image   (b) Reflection image

FIGURE 11.10:   Comparing chaotic forms: rotation-reflection

Various chaotic geometric forms appear. Special characteristics of these forms, such as the coordinates of the centre and the geometric interactions of these forms between each other, can, in some cases, be estimated. Some chaotic forms form groups or clusters in real space. Some chaotic forms are divided into other similar forms when the parameters of the model take particular values. These chaotic forms specify places in space that attract masses (particles) from the space around.

This kind of chaotic simulation of galaxy forms has the advantage that it needs limited computing power, and it mainly models "space," rather than "elements" in that space. The explanatory ability of chaotic models is high and could be useful in many cases.

11.5   Relativity in Rotation-Translation Systems

In rotation-translation systems, a number of particles follow circular paths, while, at the same time, a translation movement takes place. Numerous chaotic and non-chaotic forms arise. As we derived earlier, a two-dimensional rotation map is area preserving if the Jacobian determinant of the map is $J = 1$. The question we would like to address in this section is how the shape of the attractors changes when the speed of the rotating system relative to another system, *i.e.* the relativistic speed, influences the system parameters. In other words, how do the equations of special relativity enter into the rotation-translation equation system? The introduction of the relativity equations into a rotation-translation map will change the Jacobian of the system, and can thus produce an area-contracting map. The usual obstacle in applications is that high relativistic speeds are not possible in real situations. However, computer simulations show that even low relativistic speeds have a considerable influence on the chaotic forms.

A simple example is presented in the following figures. Figure 11.11 illustrates the chaotic image from the rotation-translation map without entering relativistic equations. A chaotic bulge appears.

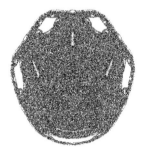

FIGURE 11.11: A rotation-translation chaotic image

The rotation-translation chaotic bulge presented in Figure 11.11 is formed when the space parameter is $b = 1$. The bulge is symmetric. The translation parameter is $a = \frac{\pi}{6}$: The other parameters are: $c = \frac{\pi}{12}$ and $d = \frac{\pi}{3}$. The equation for the rotation angle is:

$$\theta_n = c + \frac{d}{r_n}$$

where $r_n = \sqrt{x_n^2 + y_n^2}$.

The relativistic-like approach to the rotation-translation equations takes into account the length contraction governed by the equations:

$$x_{\text{rel}} = x\sqrt{1 - \frac{v^2}{c^2}}$$

$$y_{\text{rel}} = y\sqrt{1 - \frac{v^2}{c^2}}$$

where v is the relativistic speed and c is the speed of light.

Inserting these relativistic forms into the rotation-translation equations

$$\begin{aligned} x_{n+1} &= a + b(x_n \cos 2\theta_n - y_n \sin 2\theta_n) \\ y_{n+1} &= b(y_n \sin 2\theta_n + y_n \cos 2\theta_n) \end{aligned} \tag{11.6}$$

a new form is obtained. After rearrangement, the relativistic-like coefficient appears as a change in the parameter b, which now becomes:

$$b^* = b\sqrt{1 - \frac{v^2}{c^2}}$$

When $v = 0$, the parameter b is set equal to 1. When $v = c$, then $b = 0$. In reality, $v \ll c$ and the values of b are close to 1. However, even for small relativistic speeds, the influence on the chaotic bulge and on the space around the bulge is quite evident. Figure 11.12(a) illustrates the chaotic bulge and the space outside the central chaotic space. We launch particles at $x = (-1.5, -1.4, \ldots, 0)$ and $y = 0.1$. The space outside the chaotic bulge is characterised by closed curves and islands.

A relativistic speed equal to $1000 km/sec$ corresponds to $b = 0.9999945$ for the space contraction parameter b. The influence of this change of parameter b in the shape of the previous object is illustrated in Figure 11.12(b). First of all, we observe that the non-chaotic lines outside the chaotic bulge disappear. On the other hand, the lines expand, covering more space and joining each other. Over time, the space outside the chaotic image is covered by particles that are directed towards the chaotic bulge. The non-chaotic islands are shorter. This is more clear in Figure 11.12(c), in which the relativistic speed is $v = 2500 km/sec$, corresponding to a shape contraction parameter $b = 0.999965$. When $v = 2760 km/sec$, a chaotic limit is present. The islands in the chaotic sea are transformed into attracting regions, as illustrated in Figure 11.12(d). The particles from the chaotic bulge are directed into these regions, and the equilibrium points in these regions become attracting sinks.

When the relativistic speed v is between $2760 km/sec$ and $89000 km/sec$, the particles are guided to the points of attraction. However, a change in the images takes place as the relativistic speed gets higher. This is shown in Figures 11.13(a) through 11.13(f). After the limiting speed $v \approx 2760 km/sec$, symmetry breaks down. This is very clear when $v = 10000 km/sec$ and $b = 0.99944$ (Figure 11.13(b)). In Figures 11.13(c) through 11.13(e), the relativistic speed takes higher values (those values being $v = 20000 km/sec$ and $b = 0.99778$, $50000 km/sec$ and $b = 0.9860$, and $70,000 km/sec$ and $b = 0.9724$ respectively), and the images take a rotation form with three main arms and a smaller one. When $v = 89,000 km/sec$ and $b = 0.954981$, the rotating image has the form of a chaotic attractor where there are no attracting points or sinks, and the particles are trapped in the space covered by the chaotic attractor. The attractor is a totally non-symmetric rotation object with two main rotation arms and a smaller one (Figure 11.13(f)). A circular disk is centred at $(x, y) = (a, 0)$.

The chaotic attractor becomes sharper for high relativistic speeds, namely values higher than $v = 89,000 km/sec$, as illustrated in the following two figures. In Figure 11.14(a) the chaotic image is simulated for $v = 120,000 km/sec$ and $b = 0.9165$, whereas in Figure 11.14(b) the relativistic speed is in the upper limit for attractor formation, at $v \approx 170000 km/sec$ ($b = 0.8239471$).

A very surprising property of the rotation-translation system is that the chaotic image produced under the relativistic influence in large relativistic speeds is present at the beginning of the process, when time t is very small, if the area contracting parameter $b = 1$. In Figures 11.15(a) through 11.15(f), illustrations of the chaotic evolution during the first time periods appear.

The particles introduced at time $t = 0$ are distributed in a small circle centred at the origin. The radius of this circle is $r_{intr} = 0.01$. At the beginning of the process, at time $t = 10$, the image is similar to that appearing in Figure 11.14(b), where the relativistic

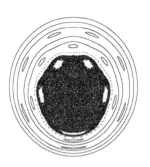

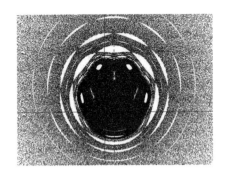

(a) Non-relativistic rotation-translation for $v = 0$ ($b = 1$)

(b) Relativistic-like rotation-translation for $v = 1000$ ($b = 0.9999945$)

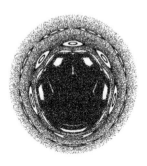

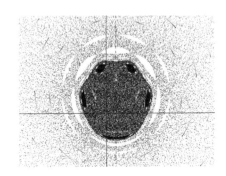

(c) Chaotic forms for speed $v = 2500 km/sec$

(d) Chaotic limit at speed $v = 2760 km/sec$

FIGURE 11.12: Chaotic forms for low relativistic speeds

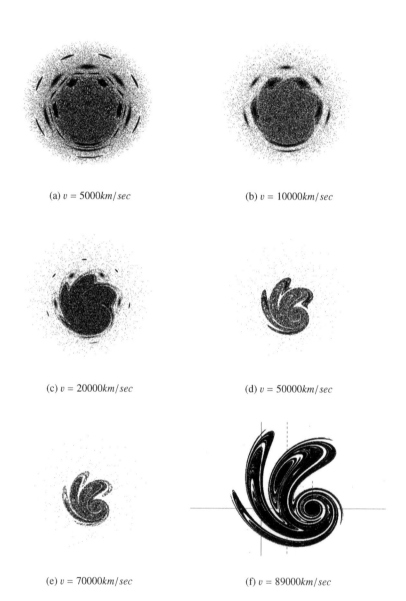

(a) $v = 5000km/sec$

(b) $v = 10000km/sec$

(c) $v = 20000km/sec$

(d) $v = 50000km/sec$

(e) $v = 70000km/sec$

(f) $v = 89000km/sec$

FIGURE 11.13: Relativistic rotation-translation forms for medium and high relativistic speeds

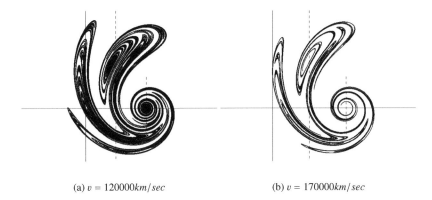

(a) $v = 120000km/sec$ (b) $v = 170000km/sec$

FIGURE 11.14: Relativistic rotation-translation forms for high relativistic speeds

speed is $v = 170000km/sec$ and the space contracting parameter is $b = 0.8239471$. Instead a similar image is now produced at time $t = 10$, but with parameter $b = 1$. The image is totally non-symmetric. As the time t increases, the chaotic images become more symmetric, and at time $t = 500$ the resulting Figure 11.15(e) is a symmetric object almost identical to the equilibrium stage image (Figure 11.15(f)) resulting when $t \to \infty$.

11.6   Other Relativistic Forms

More rotation-translation relativistic objects are based on the above equation forms but with different values for the parameters: $a = 0.8\pi$, $c = 0.4\pi$ and $d = 1.6\pi$. The resulting relativistic forms are illustrated in Figures 11.16(a) through 11.16(d).

A rotation-translation relativistic model based on the above equation forms, with parameters $a = \frac{4\pi}{3}$, $c = \frac{4\pi}{3}$ and $d = \frac{4\pi}{3}$, provides the relativistic images presented in Figures 11.17(a) and 11.17(b).

A rotation-translation relativistic object based on the above equation forms, with parameters $a = c = d = \frac{2\pi}{0.3}$, results in the relativistic forms illustrated in Figures 11.18(a) through 11.18(c).

Figure 11.18(c) illustrates a pair of galaxy-like objects formed for very high relativistic speed $v = \frac{c}{3} = 100000km/sec$. However, similar objects appear in the first time periods of the same rotation-translation procedure, but for $b = 1$, as illustrated in Figures 11.18(d) through 11.18(h). In this case, the original circular disk of particles has radius $r = 5$. Very early, at times $t = 4$ and $t = 6$, a pair of two-armed spiral galaxies appears. Later on, at time $t = 10$, a circular arm appears, and at time $t = 20$, the two spiral galaxies are connected through two connecting circular arms.

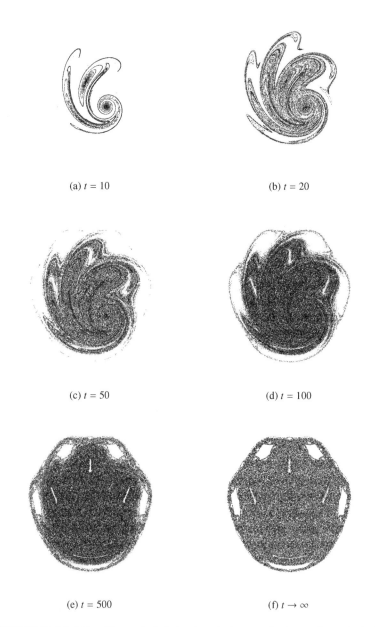

(a) $t = 10$ (b) $t = 20$

(c) $t = 50$ (d) $t = 100$

(e) $t = 500$ (f) $t \to \infty$

FIGURE 11.15:   Non-relativistic chaotic images in the early period of a
rotation-translation process with $b = 1$

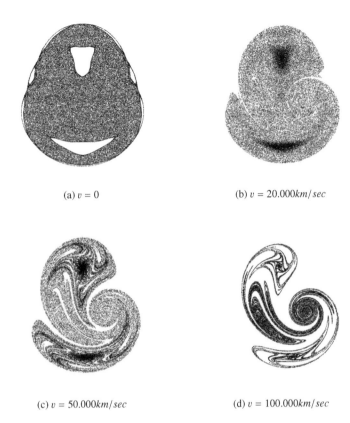

(a) $v = 0$ (b) $v = 20.000km/sec$

(c) $v = 50.000km/sec$ (d) $v = 100.000km/sec$

FIGURE 11.16: Other relativistic chaotic images

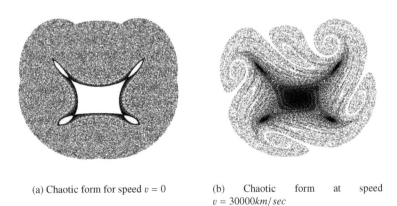

(a) Chaotic form for speed $v = 0$ (b) Chaotic form at speed $v = 30000km/sec$

FIGURE 11.17: Various relativistic chaotic forms

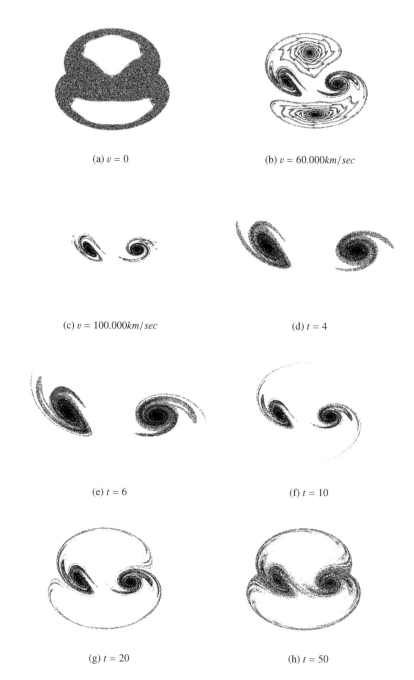

(a) $v = 0$

(b) $v = 60.000km/sec$

(c) $v = 100.000km/sec$

(d) $t = 4$

(e) $t = 6$

(f) $t = 10$

(g) $t = 20$

(h) $t = 50$

FIGURE 11.18:   Relativistic and initial chaotic images of a rotation-translation process

Even later, at time $t = 50$, the two-armed spiral galaxies are inter-connected and start to form the equilibrium shape that is illustrated in Figure 11.18(a).

Similar galaxy-like objects as above arise from the following set of parameters: $a = c = d = \frac{2\pi}{3}$. The illustrations for $b = 0.9$ are quite clear, perhaps better than in the previous case (Figure 11.19).

FIGURE 11.19:   Two galaxy-like chaotic images

Three galaxy forms are illustrated in Figure 11.20(a). These forms arise from the rotation-translation equations for $b = 0.9$. The other parameters are $a = \pi$ for the translation parameter and $c = d = \pi$ for the rotation angle parameters.

Figure 11.20(d) arises when the relativistic speed is $v = 25000 km/sec$ and $b = 0.9965217$. The three galaxy-like objects are already clearly formed. More clear forms appear for higher relativistic speeds, where $b = 0.99$ (Figure 11.20(b)). The equilibrium form object ($b = 1$) is illustrated in Figure 11.20(c).

A rotation-translation relativistic object based on the above equation forms, but with parameters $a = 1.8, b = 0.9, c = 3a$ and $d = 3a$, has the relativistic forms presented in Figure 11.21

The particles are introduced into the rotating system on the circle with radius $r = 0.1$ centred at $(x, y) = (0, 0)$. In Figure 11.21, this circle appears as a small dot in the centre of coordinates. An important point of reference is $(x, y) = (a, 0)$; and it is labelled by a cross. In the lower part of the same figure a cross indicates the point where all the particles arrive after a sufficiently large amount of time. This is simply the fixed point of the rotation-translation equations. An iterative procedure estimates this point during computer simulation.

A galaxy-like object is illustrated in Figure 11.22. The rotation-translation equations and the equation for the rotation angle are the same as above. The parameter values are $a = 5.8, c = 2.25, d = 2.5$ and $b = 0.9$. The particles are originally introduced in a circle with radius $r = 3$ and centred at the centre of coordinates (indicated by the small cross on the left of the figure). The cluster of points on the right side of the figure is centred at $(x, y) = (a, 0)$.

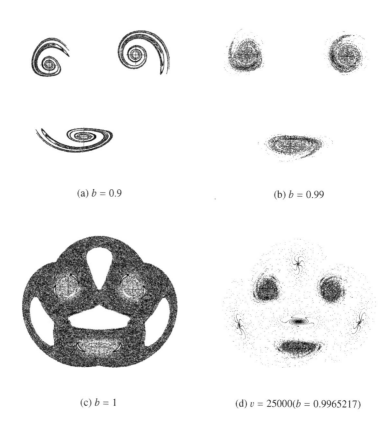

(a) $b = 0.9$ (b) $b = 0.99$

(c) $b = 1$ (d) $v = 25000(b = 0.9965217)$

FIGURE 11.20: Other relativistic chaotic images

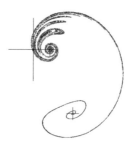

FIGURE 11.21: A relativistic chaotic image when $b = 0.9$

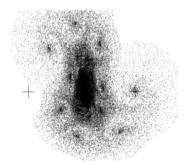

FIGURE 11.22:   A relativistic chaotic image when $b = 0.9$

11.7   Galactic Clusters

Figure 11.23 illustrates the simulation of a galactic cluster. Seven galaxy-like objects appear, while the particles originated in a disk with radius $R = 10$. A three-armed spiral is located in the middle, three two-armed spirals are connected to the three-armed spiral, and three elliptical galaxies are in the outer part. The model is a rotation-translation one expressed by the following set of equations

$$x_{n+1} = a + b(x_n \cos \theta_n - y_n \sin \theta_n)$$
$$y_{n+1} = -a + b(x_n \sin \theta_n + y_n \cos \theta_n)$$

where the rotation angle has the form

$$\theta_n = c + \frac{d}{x^2 + y^2}$$

The parameters for the simulation are $a = 2.6$, $b = 0.98$, $c = 3.2\pi$ and $d = 6$.

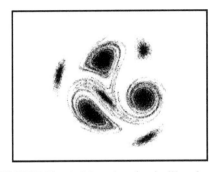

FIGURE 11.23:   A galactic-like cluster

11.8   Relativistic Reflection-Translation

The reflection-translation relativistic case is similar to the rotation-translation case, but the resulting objects are quite different. In the following, a simple example is presented. The parameter values are $a = 0.2$ for the translation parameter, and $c = 5.1$ and $d = -35$ for the reflection parameters. The reflection angle function is

$$\theta_n = c + \frac{d}{r_n^2}$$

and the reflection-translation iterative map is given by

$$x_{n+1} = a + b(x_n \cos 2\theta_n + y_n \sin 2\theta_n)$$
$$y_{n+1} = \quad b(y_n \sin 2\theta_n - y_n \cos 2\theta_n) \tag{11.7}$$

where:

$$b = \sqrt{1 - \frac{v^2}{c^2}}$$

Figure 11.24(a) illustrates the simplest case, when the relativistic speed is $v = 0$. A central circular chaotic bulge is present, located at $(x, y) = (\frac{a}{2}, 0)$. The rest of the object has a mirror-image symmetry, islands are present, as well as two characteristic points located at the solutions of $(x_{n+2}, y_{n+2}) = (x_n, y_n)$. In the case of relativistic speed, these are attracting points.

Figure 11.24(b) presents a relativistic object when the speed is much higher, $v = 1000 km/sec$. A central chaotic bulge is present, but there are two attracting points towards which the particles are directed, forming two rotating arms and creating the form of a two-armed spiral galaxy. A similar case with more clear arm structure is illustrated in Figure 11.24(c), where the relativistic speed is $v = 2000 km/sec$.

The reflection-translation at speed level $v \geq 23500 km/sec$ has special characteristics, as presented in Figure 11.24(d). The trajectories of the lower part of the plane guide the particles to an attracting point. The trajectories in the middle part of the plane guide the particles into the central bulge, whereas the trajectories of the upper part of the plane guide the particles into the two attracting arms of the galactic-like object.

11.9   Rotating Disks of Particles

11.9.1   A circular rotating disk

A number of galaxy simulations are based on the idea that particles are originally situated in a rotating disk. The simplest case is to consider a circle of some radius

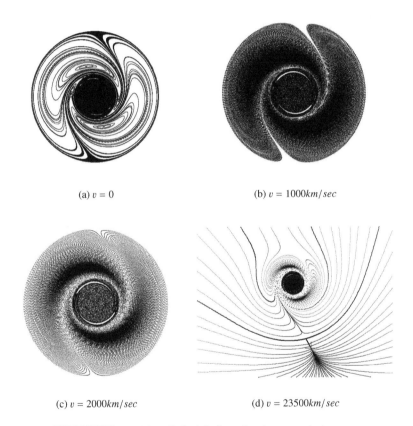

(a) $v = 0$

(b) $v = 1000 km/sec$

(c) $v = 2000 km/sec$

(d) $v = 23500 km/sec$

FIGURE 11.24: Relativistic reflection-translation

R (*i.e.* a flat disk). The particles at time $t = 0$ are distributed according to a certain distribution. For galaxy simulations, we assume that the collisions between the particles are negligible. The governing force is a central force due to gravity. In the application a cut-off radius has been used, to prevent the denominator becoming 0.

Figure 11.25(a) illustrates the rotating disk of particles in its original position. The upper left part of the figure shows the uniform distribution generating the particles, whereas the lower left part shows the density distribution of the distance of the particles from the centre. This density follows an inverse power law ($\sim 1/r^e$). The particles are placed according to a joint distribution on the polar coordinate space (r, θ), that is uniform in θ and follows an inverse power law in r ($\alpha \geq 0$):

$$f(r, \theta) = cr^{-\alpha}, \quad \theta \in [0, 2\pi], \ r \in [r_0, R] \tag{11.8}$$

Then, the density of the particles on a small annulus of radius r would follow an inverse power law ($\sim r^{-\alpha-1}$).

11.9.2   The rotating ellipsoid

Another important case is that of the formation of an ellipsoidal disk of rotating particles. In several cases, rotating bodies take the shape of a rotating ellipsoid. A disk of particles can form an ellipsoid when external forces are present. When an ellipsoidal disk is rotated, two-armed spirals appear. Similar to the aforementioned circular disk formation, an ellipsoidal disk with gravitational particles is assumed to be denser in the centre, and then the density gradually decreases from the centre to the outer part of the ellipsoid, as illustrated in Figure 11.25(b).

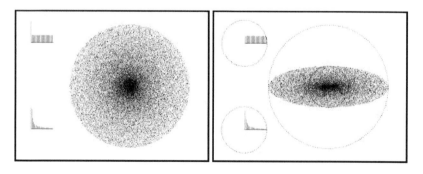

(a) A rotating circular disk   (b) A rotating ellipsoidal disk

FIGURE 11.25:   Rotating disks

The set of equations that provide the two-dimensional ellipsoid is

$$x_t = a \cos \phi_t$$
$$y_t = b \sin \phi_t \qquad (11.9)$$
$$z_t = z_0$$

or:

$$\left(\frac{x}{a}\right)^2 + \left(\frac{y}{b}\right)^2 = 1$$

When the system turns by an angle of θ_t from time t to time $t + 1$, the equations become:

$$x_{t+1} = a \cos(\phi_t + \theta_t)$$
$$y_{t+1} = b \sin(\phi_t + \theta_t) \qquad (11.10)$$
$$z_{t+1} = z_0$$

Using the trigonometric sum formulas, the following iterative formula can be obtained:

$$x_{t+1} = x_t \cos \theta_t - \frac{a}{b} y_t \sin \theta_t$$
$$y_{t+1} = \frac{b}{a} x_t \sin \theta_t + y_t \cos \theta_t \qquad (11.11)$$
$$z_{t+1} = z_0$$

By setting $h = \frac{b}{a}$, assuming that $0 < b < a$ ($0 < h < 1$), the system (11.11) takes the form:

$$x_{t+1} = x_t \cos \theta_t - \frac{1}{h} y_t \sin \theta_t$$
$$y_{t+1} = h x_t \sin \theta_t + y_t \cos \theta_t \qquad (11.12)$$
$$z_{t+1} = z_0$$

In Figure 11.26(a) a two-armed spiral galaxy is formed, based on a rotating flat ellipsoid with eccentricity parameter $h = \frac{b}{a} = \frac{1}{3}$. The system (11.12) is used, with the rotation angle being

$$\theta_t = \frac{d}{\sqrt{r_0 + (30r)^7}}$$

where the cut-off radius is $r_0 = 0.01$ and the rotation parameter is $d = 6.7$.

The elapsed time is $t = 10$ time units. When $t = 1$, the form of a two-armed spiral results (Figure 11.26(b)). By observing the two images, it becomes clear that the original two-armed spiral at time $t = 1$ is transforming into a two-armed bar formation after time $t = 10$.

The ellipsoidal disk of particles is rotated counterclockwise. After $t = 5$ time units, a two-armed spiral is present, located at the centre of coordinates (Figure 11.27(a)). The rotation follows an angle equal to $\theta = \frac{0.01}{0.001 + r^2}$, whereas $b = 1$, resulting in an area-preserving map. A central chaotic bulge leads to two spiral arms.

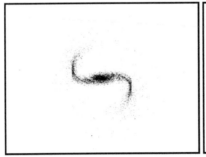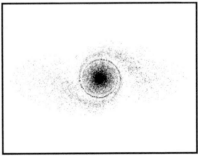

(a) A two-armed bar galaxy ($t = 10$)   (b) A two-armed spiral galaxy ($t = 1$)

FIGURE 11.26:   Two-armed galaxies

(a) A two-armed spiral galaxy ($t = 5$)   (b) A two-armed spiral galaxy ($t = 5$)
                                                 with space contraction $b = 0.9$

FIGURE 11.27:   Two-armed galaxies

A similar case, but with $b = 0.9$, is presented in Figure 11.27(b). After time $t = 5$, the original ellipsoid is smaller and the resulting two-armed spiral has a central region resembling a bar, rather than a simple bulge. The space contraction relations for the ellipsoid are

$$x_{t+1} = b\left(x_t \cos \theta_t - \frac{1}{h} y_t \sin \theta_t \right)$$

$$y_{t+1} = b\left(h x_t \sin \theta_t + y_t \cos \theta_t \right) \tag{11.13}$$

$$z_{t+1} = z_0$$

11.10   Rotating Particles under Distant Attracting Masses

11.10.1   One attracting mass

We consider a number of rotating particles in a disk of radius $r = 1$, attracted by an outside mass located on the x-axis, at distance h from the origin. The density of the disk of the initial rotating particles follows an inverse power law as previously. The particles rotate with the same angular velocity. The issue at hand is to explore the shape of this disk of particles after the outside mass has exerted its influence for time t. The attracting force is of course the gravitational force. If the rotation velocity of the particles on the disk is very small, or equal to zero, the particles will be attracted by the mass and will be directed to it. But, as the angular velocity is higher, a number of particles are trapped inside the rotating disk. The distribution of these particles inside the disk is not uniform. A central chaotic region can be seen. The paths of the particles are chaotic in this region. The particles are trapped in this chaotic region, and aside from their chaotic movement, they tend to keep their position in the chaotic region. In other words the chaotic movement is responsible for a *space stability*, in that the particles move chaotically in the fixed space. On the other hand, the particles moving outside this region are following non-chaotic paths, and are mainly attracted by the outside mass.

The rotation-attraction model results from the rotation model after a transformation that takes into account the influence of the attracting mass. The two-dimensional equations have the form

$$x_{t+1} = \frac{d(h - x)}{((h - x)^2 + y^2)^{3/2}} + x_t \cos \theta_t - y_t \sin \theta_t$$

$$y_{t+1} = -\frac{dy}{((h - x)^2 + y^2)^{3/2}} + x_t \sin \theta_t + y_t \cos \theta_t \tag{11.14}$$

where h and d are parameters and the rotation angle is:

$$\theta_t = \frac{c}{(r_0^2 + r_t^2)^{3/2}}$$

The cut-off radius is r_0, c is a rotation parameter and $r_t = \sqrt{x_t^2 + y_t^2}$. As is obvious from (11.14), the influence of the attracting mass is expressed by the first term on the right hand side of the relations. These terms arise by considering the gravitational force acting on the small mass m of a rotating particle located at (x, y), from the attracting mass M located on the x-axis at $(x, y) = (h, 0)$. Therefore the Euclidean distance between the particle m and the mass M is

$$r_{(m,M)} = \sqrt{(h - x)^2 + y^2}$$

and the attracting force is

$$F = \frac{GMm}{r_{(m,M)}^2}$$

where G is the gravitational constant. The acceleration of the particle therefore is:

$$\gamma_m = \frac{GM}{r_{(m,M)}^2}$$

The x-component of the acceleration is

$$\gamma_x = \frac{\gamma_m(h - x)}{r_{(m,M)}} = \frac{GM(h - x)}{r_{(m,M)}^3}$$

or

$$\gamma_x = \frac{d(h - x)}{((h - x)^2 + y^2)^{3/2}}$$

where $d = GM$.

Similarly, the y-component of the acceleration is

$$\gamma_y = -\frac{\gamma_m y}{r_{(m,M)}} = -\frac{GMy}{r_{(m,M)}^3}$$

or:

$$\gamma_y = -\frac{dy}{((h - x)^2 + y^2)^{3/2}}$$

In other words, the x-component of the acceleration moves the particle towards the positive x-direction, whereas the y-component of the acceleration directs the particle towards the centre of the rotating disk.

Figure 11.28(a) illustrates the rotating disk at time $t = 0$, before the attracting mass is introduced. Small circles in the periphery, at $r = 1$, indicate a number of rotating particles. Figure 11.28(b) shows the effect of the introduction of an attracting mass at distance $h = 3$ from the origin. The parameters are $r_0 = 0.001$, $c = 0.4$ and $d = 0.8$, and the elapsed time is $t = 5$.

In Figure 11.28(b), the original outer (peripheral) particles change from a circular form into an egg-shaped image. However, several particles that originated from

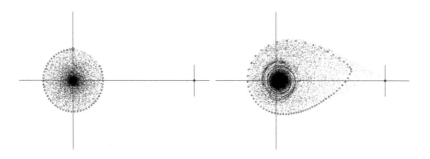

(a) The rotating disk, $r = 1$. Outer particles at $r = 1$ rotate before the attracting
mass is introduced.

(b) The rotating particles after time $t =$
5 from the introduction of the attracting
mass

FIGURE 11.28: Rotating particles under the influence of an attracting mass

within the disc, with $r < 1$, are attracted from the mass M, escape outside the egg-shaped form and approach the attracting mass. These particles have an energy level higher than the escape energy level. The chaotic part of this figure is located in the central area of the rotating disk. The particles in the chaotic region at time $t = 5$ form a two-armed spiral rotating image. A chaotic path of such a particle is presented in Figure 11.29(a). The particle follows a box-like path. The elapsed time is $t = 100$ and the parameters are $c = 0.2$, $d = 0.3$. The original position of the particle is at $(x, y) = (0.2, 0.6)$.

Figure 11.29(b) illustrates a view of the two-armed spiral galaxy and the surrounding cloud at time $t = 4$. The original disk radius is $r = 1$, the cut-off radius is $r_0 = 0.001$, and the density follows an inverse power law ($\sim 1/r^5$). The parameters are $c = 0.567$, $d = 0.1$, and the distance of the attracting mass is $h = 3$.

11.10.2 The area of the chaotic region in galaxy simulations

Computer experiments show that the chaotic region in the above simulations is larger when the rotation velocity is higher. However, it is hard to say that the construction of a chaotic area equation is possible. Another important point is that the chaotic area changes as the time t varies. If the chaotic area is stabilised after enough time has elapsed, then we say that the stochastic area is *asymptotically stable*, and we can search for possible estimation methods. One approach is to construct a multidimensional graph related to the chaotic area, with the rotation velocity, the attracting mass M and the distance of the attracting mass h as parameters.

Such an asymptotically stable chaotic attractor is illustrated in Figure 11.30. The circular central part and the rotating arms appear. The attracting mass M is located at distance $h = 3$ to the right of the attractor, at $(x, y) = (3, 0)$. The parameters are $c = 0.5$ and $d = 1$. The centre of the circular central part has moved to the right, at

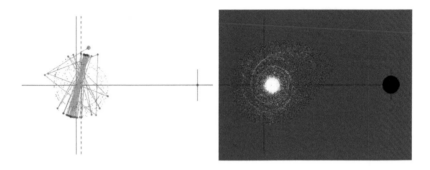

(a) A particle trapped in the chaotic region, that oscillates between the spiral arms

(b) A view of the two-armed spiral galaxy and the surrounding cloud at time $t = 4$

FIGURE 11.29: Box-like paths of a two-armed spiral galaxy

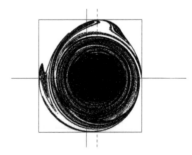

FIGURE 11.30: An asymptotically stable chaotic attractor

$(x, y) = (0.117, 0)$. The outer parts of the attractor extend to $x \approx 0.61$ to the right and $x \approx -0.52$ to the left. The larger y coordinate is at $y \approx 0.62$, and the smaller is at $y \approx -0.57$.

11.10.3   The speed of particles

There are several theories for the distribution of the speed of particles in rotating disks. Furthermore, observations of galaxies show that the mean rotation speed in the periphery of the disk is higher than the speed predicted from theory. This urged astronomers to conclude the existence of non-observed mass in the galaxies, the well-known *dark matter*. Every simulation based on a rotating disk of particles has to take into account the gradual change of the particle density, an inverse power law of $\sim 1/r^5$ seems to be acceptable, and results in a rotation velocity distribution that is closer to the results of dark matter theory. As it is not, even nowadays, completely clear how to find a global model for the rotating disks of particles that would be quite close to real situations, a test of our model regarding the rotation speed is called for.

The simple assumption of a stable angular rotation speed will lead to very high rotation speeds for large r. In the proposed models, the rotation angle is usually equal to an inverse function of the radius r from the origin, giving a function for the rotation speed that is close to some real situations. In the following computer experiment, the rotation speed is estimated and presented in Figure 11.31.

The rotation disk has radius $r = 2$. The rotation equations (11.14) are those presented above. Estimates of the mean speed of particles at concentric areas of radius differing by 0.1 are presented in Figure 11.31. The graph verifies the results known from real situations. The rotation speed is initially growing and reaching an upper limit; thereafter it gradually declines. A divergence is estimated at radius $r = 0.40$ and $r = 0.45$.

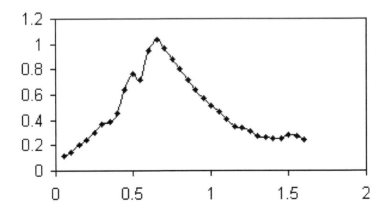

FIGURE 11.31:   Rotation speed vs. radius r

11.11 Two Equal Attracting Masses in Opposite Directions

We now consider the case where the rotating particles in a disk of radius $r = 1$ are attracted by two equal outside masses, located on the x-axis at distances h and $-h$ from the origin. The density of the disk of rotating particles follows an inverse power law, of the order $1/r^n$, where $n = 3, 4, 5, \ldots$. The particles rotate with the same angular velocity. As before, we wish to explore the shape of the original disk of particles after the outside masses have exerted their influence for time t. The attracting force is again the gravitational force.

The situation is in many ways similar to that of one attracting mass. If the rotation velocity of the particles on the disk is very small, or equal to zero, the particles on the disk will be attracted from the masses and will be separated in two parts and directed towards the masses. On the other hand, when the angular velocity is higher, a number of particles are trapped inside the rotating disk. The space inside this disk is not uniform. A central chaotic region is again visible. The paths of the particles are chaotic in this region, exhibiting the same kind of space stability as before. The particles moving outside this region are following non-chaotic paths, mainly influenced and attracted by the outside masses. As the two equal masses act in opposite directions, the resulting forms will be symmetric and the galactic forms would be two-armed spirals.

The two-mass-rotation-attraction model results from (11.14) by taking into account the influence of the two opposite attracting masses. The two-dimensional equations have the form

$$
\begin{aligned}
x_{t+1} &= A(h - x_t) + x_t \cos\theta_t - y_t \sin\theta_t \\
y_{t+1} &= -Ay_t \qquad\quad + x_t \sin\theta_t + y_t \cos\theta_t
\end{aligned}
\tag{11.15}
$$

where A is given by

$$
A = \frac{d}{((h - x)^2 + y^2)^{3/2}} + \frac{d}{((-h - x)^2 + y^2)^{3/2}}
$$

h and d are parameters, and the rotation angle is:

$$
\theta_t = \frac{c}{(r_0^2 + r_t^2)^{3/2}}
$$

The cut-off radius is r_0, c is a rotation parameter and $r_t = \sqrt{x_t^2 + y_t^2}$.

As is clear from the above map, the influence of the attracting masses is expressed by the term A. This term arises by considering the gravitational force acting on the small mass m of a rotating particle located at (x, y) from the attracting masses M located at $(x, y) = (h, 0)$ and $(x, y) = (-h, 0)$ on the x-axis, similar to the case of one

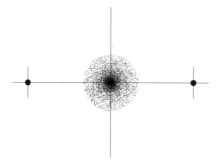

FIGURE 11.32: Two attracting masses and the disk of rotating particles at time $t = 0$

attracting mass. Figure 11.32 illustrates this situation at time $t = 0$. The attracting masses are located on the x-axis at distances $h_1 = 3$ and $h_2 = -3$ from the origin.

Figure 11.33(a) shows the results of the simulation. The parameters are $d = GM$ and $h = 3$. The resulting rotation-attraction form after time $t = 1$ is a perfect symmetric two-armed spiral.

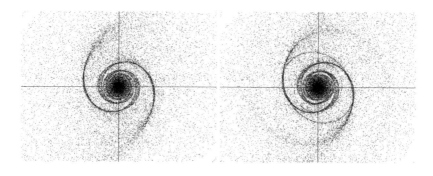

(a) A two-armed spiral after time $t = 1$ (b) Rotation-attraction image at $t = 2$

FIGURE 11.33: Rotating disks with two attracting masses

At time $t = 2$, the two-armed spiral starts to change to a more complicated form (Figure 11.33(b)), whereas at time $t = 5$ the resulting form is closer to an ellipsoidal centre with outer circular rings (Figure 11.34(a)). The asymptotically stable rotation-attraction form for large time t is illustrated in Figure 11.34(b).

Another example is related to two equal masses located on the x-axis, at $h_1 = 3$ and $h_2 = -3$, and influencing the original rotating disk of particles. The parameters are $c = 2.2$ and $d = d_1 = 2$. For sufficiently large time t, the circular rotating disk takes the form presented in Figure 11.34(c). The centre of the circle-like central

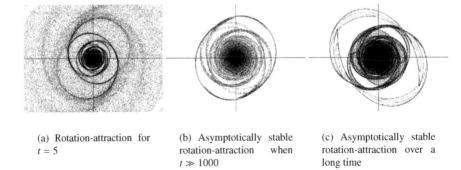

(a) Rotation-attraction for $t = 5$

(b) Asymptotically stable rotation-attraction when $t \gg 1000$

(c) Asymptotically stable rotation-attraction over a long time

FIGURE 11.34:   Rotating disks with two attracting masses over a long time

disk (bulge) is at the centre of coordinates. Symmetric arms and eye-like forms are extended close to the central disk.

11.11.1   Symmetric unequal attracting masses

In the above example the two equal attracting masses act from opposite directions in the x-axis on the rotating disk of particles. The centre of this disk was originally located at the centre of coordinates. In this section we examine the case where the two masses located at $(x, y) = (h = 3, 0)$ and $(x, y) = (h1 = -3, 0)$ have unequal masses. Then, equations (11.15) still hold, where A now is:

$$A = \frac{d_1}{((h-x)^2 + y^2)^{3/2}} + \frac{d_2}{((-h-x)^2 + y^2)^{3/2}}$$

We will take those masses to be M for the first and $M/4$ for the second. Now, as the opposing attracting forces are not equal, the disk of particles moves to the right and the galaxy arms are not symmetric. In Figure 11.35(a) the parameters are $c = 0.5$, $d_1 = 2$ for the large mass M and $d_2 = d_1/4$ for the small mass $M/4$. The object is a chaotic attractor. This is the attracting space for every original point mass left in the vicinity of the attractor and having enough time to enter the attracting space. Two main arms appear, as well as a smaller third arm originating from the second arm.

Figure 11.35(b) illustrates the case where the parameter expressing rotation has a higher value, $c = 0.8$. Two non-symmetric arms appear. These are connected to the central disk with smaller circular parts. The effect of a varying parameter c is illustrated in Figures 11.35(c) and 11.35(d). The chaotic arms are larger, as the parameter is $c = 1.1$ in the first figure and $c = 2$ in the second.

Another interesting point is the form that the initial disk takes at the very early time periods. The parameters are $c = 2$, $d_1 = 2$, $d_2 = d_1/4$. At time $t = 1$, a two-armed spiral is formed (Figure 11.36(a)). The two arms are different. The larger one is directed towards the bigger attracting mass to the right of the rotating disk, while

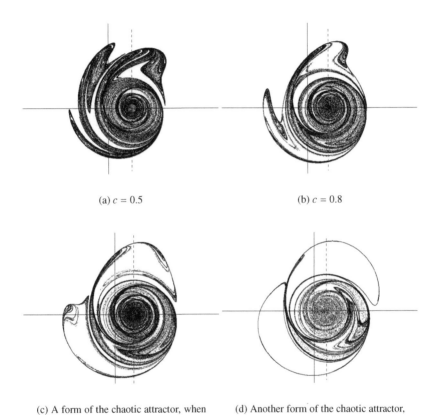

(a) $c = 0.5$

(b) $c = 0.8$

(c) A form of the chaotic attractor, when $c = 1.1$

(d) Another form of the chaotic attractor, when $c = 2$

FIGURE 11.35: Chaotic attractors in the case of two unequal symmetric attracting masses

the smaller is directed towards the small attracting mass to the left of the centre of coordinates.

At time $t = 2$, both spiral arms are directed to the big attracting mass on the right of the coordinate centre (Figure 11.36(b)). At time $t = 3$ a new spiral arm is added (Figure 11.36(c)). All three arms are directed towards the big attracting mass located on the right-hand side. At time $t = 4$ the total number of the spiral arms is now 4 and all are directed towards the big attracting mass on the right (Figure 11.36(d)). More complicated forms arise at times $t = 5$, $t = 6$ and $t = 8$ as illustrated in Figures 11.37(a) through 11.37(c).

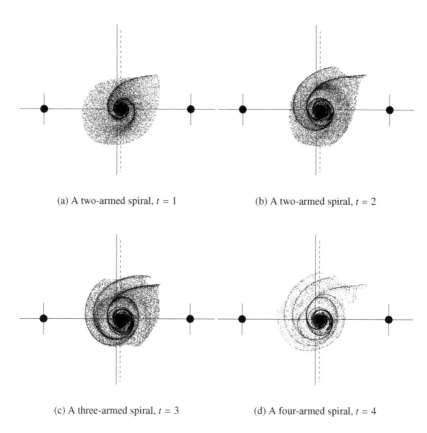

(a) A two-armed spiral, $t = 1$ (b) A two-armed spiral, $t = 2$

(c) A three-armed spiral, $t = 3$ (d) A four-armed spiral, $t = 4$

FIGURE 11.36: Rotating disc at early times

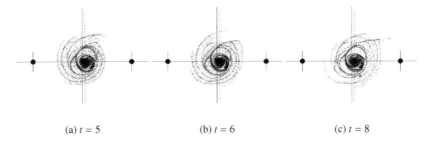

(a) $t = 5$ (b) $t = 6$ (c) $t = 8$

FIGURE 11.37: Multi-arm images

11.12 Two Attracting Equal Nonsymmetric Masses

We consider now the case where there are two equal attracting masses, located on the x-axis and y-axis respectively, at equal distance h from the origin. Once again, the density of the disk of rotating particles follows an inverse power law. As in the previous cases, the space inside this disk is not uniform. A central chaotic region is again visible, and paths of the particles in this region are chaotic. As the two equal masses act on the rotating particles from perpendicular directions, the resulting forms must be non-symmetric, and the galactic forms would be non-symmetric two-armed spirals.

The rotation-attraction model in this case takes the form

$$\begin{aligned} x_{t+1} &= A_1(h - x) - A_2 y + x_t \cos \theta_t - y_t \sin \theta_t \\ y_{t+1} &= -A_1 y + A_2(h - y) + x_t \sin \theta_t + y_t \cos \theta_t \end{aligned} \tag{11.16}$$

where

$$A_1 = \frac{d}{((h - x)^2 + y^2)^{3/2}}$$

$$A_2 = \frac{d}{(x^2 + (h - y)^2)^{3/2}}$$

h and d are parameters, and the rotation angle is

$$\theta_t = \frac{c}{(r_0^2 + r_t^2)^{3/2}}$$

r_0 is the cut-off radius, c is a rotation parameter and $r_t = \sqrt{x_t^2 + y_t^2}$.

As is obvious from the above map, the influence of the attracting masses is expressed by terms A_1, A_2 respectively. These terms arise by similar considerations as above. Figure 11.38(a) illustrates the situation at time $t = 1$. The distance is $h = 3$, and the rotating disk radius is $r = 1.5$.

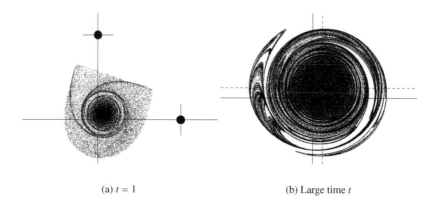

(a) $t = 1$ (b) Large time t

FIGURE 11.38: The effect of two perpendicularly attracting masses

The two spiral arms are directed towards the attracting masses. The central bulge has a circular form. Another circular ring outside the central bulge is the starting point of the spiral arms. The parameters are $c = 1.3$, $d = 1$ and $h = 3$. Two non-symmetric spiral arms are formed for large time t as presented in Figure 11.38(b).

Questions and Exercises

1. Show that equations (11.1) imply the following relation for the motion of a mass m under the influence of one large mass M, with $MG = 1$:

$$\frac{1}{2}\left(x'^2 + y'^2\right) - \frac{1}{r} = \text{const.}$$

2. From the relation in the previous exercise, determine the escape velocity, by considering that for a mass to escape, its speed at $r = \infty$ should be "non-negative."

3. Letting $MG = 1$ for simplicity, the equation

$$\frac{mv_{\text{cycl}}^2}{r} = \frac{mMG}{r^2} = \frac{m}{r^2}$$

gives us the velocity of a particle that moves in a circular orbit of radius r around an attracting mass M. Use this equation, in combination with the results in the previous questions, to show that the escape velocity of a particle at some distance from the mass, and the rotating velocity of that particle at the same distance, are related via:

$$v_{\text{esc}} = \sqrt{2}v_{\text{cycl}}$$

4. Consider the system

$$x_1' = x_3$$
$$x_2' = x_4$$
$$x_3' = -\frac{1}{r^c}x_1$$
$$x_4' = -\frac{1}{r^c}x_2$$

where $r = \sqrt{x_1^2 + x_2^2}$. Find the trajectories of this system in the (x_1, x_2) plane when $c = \sqrt{2}$. Do the same when $c = 1$ and $c = 1/2$. (Use a fourth order Runge-Kutta algorithm and initial values $x_1 = 1$, $x_2 = 0$, $x_3 = 0$ and $x_4 = 1.3$.)

5. Use the translation reflection equations

$$x_{t+1} = a + x_t \cos(2\theta_t) + y_t \sin(2\theta_t)$$
$$y_{t+1} = \qquad x_t \sin(2\theta_t) - y_t \cos(2\theta_t)$$

to create images with outer and inner formations. The rotation angle should have the form $\theta_t = c + d/r$. One interesting choice of parameters is: $a = 1$, $c = 5.1$ and $d = -20$, with initial values $x = a$ and $y = 0$. Try other parameter values as well.

6. Use the parametric equations

$$x_t = r_t \cos(t)$$
$$y_t = r_t \sin(t)$$

where

$$r_t = \frac{ep}{1 - e\cos(t)}$$

to obtain ellipses. Start with parameter values $e = 0.5$ and $p = 2$.

7. Use the parametric equations

$$x_t = r_t \cos(tr_t)$$
$$y_t = r_t \sin(tr_t)$$

where

$$r_t = \frac{ep}{1 - e\cos(t)}$$

to draw trajectories in the (x_t, y_t) plane. Start with parameter values $e = 0.5$ and $p = 2$, and try several values for e. What do you observe?

8. Draw the (x_t, y_t) trajectories for the iterative process:

$$x_{t+1} = 0.15 + 0.5r_t \cos(t)$$
$$y_{t+1} = \qquad\qquad r_t \sin(t)$$

where

$$r_t = 0.5 - \frac{1}{\sqrt{x_t^2 + y_t^2}}$$

9. For the Contopoulos-Bozis simulation described in Section 11.3.1, try other values for the space contracting parameter b and observe the resulting (x_t, y_t) diagrams.

10. Show that if particles are distributed on a disk according to the probability density function (11.8), then the density of particles on any thin annulus at radius r will be proportional to $r^{-\alpha-1}$.

Chapter 12

Galactic-Type Potentials and the Hénon-Heiles System

12.1 Introduction

The work of Hénon and Heiles on galactic motion (Hénon and Heiles, 1964) is the first complete work proving the existence of chaos in galaxy formations. From its first appearance in the *Astronomical Journal* (1964), it provided considerable evidence of the existence of chaotic motions in galaxies. The work by Hénon and Heiles started as a computer experiment in order to explore the existence and applicability of the third integral of galactic motion. Earlier, in 1956, George Contopoulos (Contopoulos, 1956) started a pioneering work on galactic motion and renewed the interest on the third integral of motion (Contopoulos, 1960). Contopoulos explored the box-like paths of a body in a galactic potential and found that the orbits in the meridian plane of an axisymmetric galaxy are like Lissajous[1] figures (Contopoulos, 1965). Contopoulos expected the orbits to be ergodic and fill all the space inside the energy surface (Contopoulos, 2002). Instead, he found that the orbits did not fill all the available space, but instead filled curvilinear parallelograms, like deformed Lissajous figures. According to his writings he could prove later (Contopoulos, 1960) that such orbits can be explained qualitatively and quantitatively by a formal third integral of motion. The work of Contopoulos and Hénon-Heiles was a result of what we call *computer experiments*. This new type of experiments gave new directions to various scientific fields, including astronomy. The computer results were sometimes surprising, and often contradicting existing theories. Contopoulos (1958) used a computer in Stockholm Observatory in 1958, and Hénon and Heiles (1964) used a computer at Princeton University from 1962 to 1963.

---

[1]*Lissajous curves* are given in parametric form by equations that describe complex harmonic motion.

12.2 The Hénon-Heiles System

The Hamiltonian applied by Hénon and Heiles (1964) is of the form:

$$H = \frac{1}{2}(\dot{x}^2 + \dot{y}^2) + \frac{1}{2}\left(x^2 + y^2 + 2x^2y - \frac{2y^3}{3}\right) = h \tag{12.1}$$

where the *potential* $U(x, y)$ is

$$U(x, y) = \frac{1}{2}\left(x^2 + y^2 + 2x^2y - \frac{2y^3}{3}\right) \tag{12.2}$$

or in polar coordinates:

$$V(r, \theta) = \frac{1}{2}r^2 + \frac{1}{3}r^3 \sin(3\theta)$$

A more general form for a potential would be:

$$V(r, \theta) = r^4 + ar^2 + br^3 \cos(3\theta)$$

The Hamiltonian for a generalised Hénon-Heiles potential is ($p_x = \dot{x}, p_y = \dot{y}$)

$$H = \frac{1}{2}(p_x^2 + p_y^2 + Ax^2 + By^2) + Dx^2y - \frac{1}{3}Cy^3$$

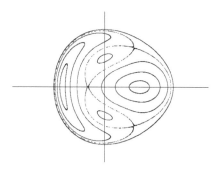

FIGURE 12.1: The original Hénon-Heiles $(y, \dot{y})$ diagram

Here, we reproduce by simulation the original figure presented in the Hénon-Heiles paper (Figure 12.1). They estimated the paths of a particle according to the previously mentioned Hamiltonian at an energy level $E = h = 1/12$, and then they gave the $(y, \dot{y})$ diagram when $x = 0$. At this low energy level mainly regular orbits appear. At a higher energy level chaotic regions are formed.

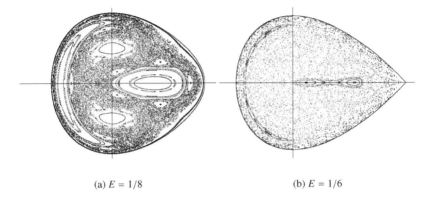

(a) $E = 1/8$ (b) $E = 1/6$

FIGURE 12.2: $(y, \dot{y})$ diagrams for the Hénon-Heiles system

A Poincaré section at $x = 0$ is presented in Figure 12.2(a) for the Hénon-Heiles model. The energy level is at $E = 1/8 = 0.125$. Regular and chaotic trajectories are present. Several islands appear surrounded by the chaotic sea.

In the Hénon-Heiles Hamiltonian, the upper limit for the energy is $E = h = 1/6$. Figure 12.2(b) illustrates this case. The $(y, \dot{y})$ diagram is almost totally chaotic. However, small islands remain inside the chaotic sea.

The Hénon-Heiles system is given by the following set of differential equations:

$$
\begin{aligned}
\dot{x} &= u \\
\dot{y} &= v \\
\dot{u} &= -x - 2xy \\
\dot{v} &= -y - x^2 + y^2
\end{aligned}
\tag{12.3}
$$

The Jacobian of this system is

$$ J = 1 - 4\left(x^2 + y^2\right) = 1 - 4r^2 $$

This system results from the Hamiltonian (12.1), and the related potential (12.2) that Hénon and Heiles chose for their famous computer experiment. The selection of this function to express the potential in the above Hamiltonian has the advantages that it is relatively simple but, at the same time, it gives adequate information related to chaotic and non-chaotic paths. During those early days of computer use and the discovery of chaotic surprises in non-linear systems, it was important to use simple non-linear functions in order to explore chaos. Many scientists of the time were able to verify the validity of the theory and expand the results into new fields. Contopoulos (1960) explored a field of non-linear dynamics in a galaxy, which was later analysed at Princeton University in 1963 by the French astronomer Michel Hénon and his colleague Carl Heiles from the United States. The paper was published one year later in the *Astronomical Journal* (Hénon and Heiles, 1964). In 1963 Lorenz's famous work on chaotic modelling was also published (Lorenz, 1963).

12.3   Discrete Analogues to the Hénon-Heiles System

In this section we examine discrete analogues to the Hénon-Heiles model. Discrete models are very important in studying chaotic systems, as they are usually simpler than the continuous time systems and, quite often, they explore other aspects of the chaotic system that are not present in continuous time models. The simulation of this discrete model gives an image presented in Figure 12.3 at the energy level $E = 1/12$. The parameter c was set to 0.1. The discrete model is based on the following set of equations:

$$x_{n+1} = cx_n + u_n$$
$$y_{n+1} = cy_n + v_n$$
$$u_n = -x_n - 2x_{n+1}y_{n+1} \qquad (12.4)$$
$$v_n = -y_n - x_{n+1}^2 + y_{n+1}^2$$

The Jacobian of the discrete model is:

$$J = 1 + 2(c - 1)y_{n+1} + 4(c - 1)\left(x_{n+1}^2 + y_{n+1}^2\right)$$

It is obvious that, when $c = 1$, $J = 1$ and then the space-preserving property is sat-

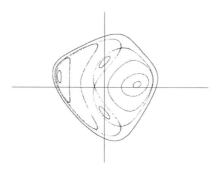

FIGURE 12.3:   A discrete analogue to the Hénon-Heiles system ($E = 1/12$)

isfied. However, when c takes values less than 1, very interesting behaviour results, as in Figure 12.3, where $c = 0.1$. The similarities with the Hénon-Heiles system are obvious when comparing Figure 12.3 with Figure 12.1.

Hénon and Heiles proposed and applied a different, simple, discrete model:

$$x_{n+1} = x_n + a\left(y_n - y_n^3\right)$$
$$y_{n+1} = y_n - a\left(x_{n+1} - x_{n+1}^3\right) \qquad (12.5)$$

This model has only one parameter a. When $a = 1.6$, Figure 12.4(a) results from a set of initial values (x, y). A very interesting property of model (12.5) arises when $a = 0.1$. The form presented in Figure 12.4(b) has similarities with the above proposed more complicated discrete form of the Hénon-Heiles model.

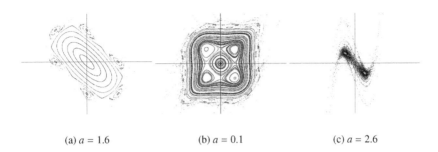

(a) $a = 1.6$ (b) $a = 0.1$ (c) $a = 2.6$

FIGURE 12.4: The discrete Hénon-Heiles model

The simple model used by Hénon and Heiles has more interesting properties as is illustrated in Figure 12.4(c). Here, $a = 2.6$, and the graph represents a bar-like image with two tails. The challenge now is to apply rotation to this simple model, in order to explore the new features of the model. The classical rotation equation is applied, with a stable rotation angle $\theta = 0.9$. The graph, in this case, turns out to be a chaotic attractor, as illustrated in Figure 12.5(a).

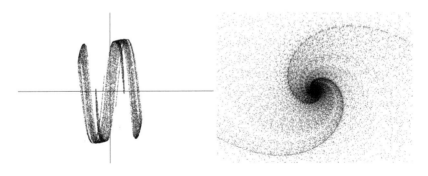

(a) A chaotic attractor for the discrete rotation Hénon-Heiles model   (b) A two-armed spiral galaxy

FIGURE 12.5: Discrete analogues to the Hénon-Heiles system

The graph in Figure 12.5(b), illustrating a two-armed spiral galaxy, is produced by using the following simple rotation model:

$$x_{n+1} = b(x_n \cos a - y_n \sin a) + x_n^2 \sin a$$
$$y_{n+1} = b(x_n \sin a + y_n \cos a) - x_n^2 \cos a$$

The parameter values for the simulation were $a = 0.2$ and $b = 0.9$. As the Jacobian is $J = b^2$, the second parameter is an area-contracting parameter.

12.4   Paths of Particles in the Hénon-Heiles System

The Hénon-Heiles system examined above also gives rise to some interesting paths in the (x, y) plane. The paths are characterised by the level of the potential at that stage. The form of the potential is triangular. A totally chaotic trajectory is illustrated in Figures 12.6(a) and 12.6(b). The energy level is $E = h = 1/6$. This is the escape limit, at which the equipotential triangle is drawn. All the paths of the particle are included within the triangle. The parameters in this case were $x = 0$, $y = 0$ and $v = 0.1$. The value of u, $u = 0.5668627\cdots$, was determined by the condition that the energy level of Hamiltonian should be $E = h = 1/6$.

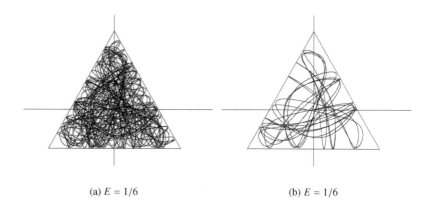

(a) $E = 1/6$ (b) $E = 1/6$

FIGURE 12.6:   Chaotic paths in the Hénon-Heiles system

Figure 12.7(a) illustrates the equipotential lines for the Hénon-Heiles system. The particle is trapped inside this space, and the paths follow ordinary or chaotic paths depending on the energy level in each case. The limits for the potential are estimated by computing the critical points of the potential function, solving the system:

$$\frac{\partial U}{\partial x} = -\ddot{x} = x + 2xy = 0$$

$$\frac{\partial U}{\partial y} = -\ddot{y} = y + x^2 - y^2 = 0$$

This system has four solutions. One is $(0,0)$, the point with the minimum potential. The other three are all points of maximum potential $U(x,y) = 1/6$:

$$(x,y) = (0,1)$$
$$(x,y) = (\ \sqrt{3}/2, -1/2)$$
$$(x,y) = (-\sqrt{3}/2, -1/2)$$

It is, then, easy to draw the equilateral triangle with corners located at these three coordinates for (x,y). In Figure 12.7(a), several equipotential lines are drawn. The potential U is constrained in the interval $(0, 1/6)$. A non-chaotic orbit is computed and presented in Figure 12.7(b) for the energy level $E = 1/12$. The parameter values for the simulation were $x = 0.39581$, $y = 0$, $\dot{x} = u = 0.1$ and $\dot{y} = v = 0$.

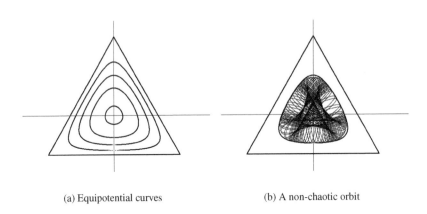

(a) Equipotential curves   (b) A non-chaotic orbit

FIGURE 12.7: The Hénon-Heiles system

12.5 Other Forms for the Hamiltonian

Several other forms for the Hamiltonian can be used in galaxy simulations. A number of these models are based on Hamiltonians with a fourth power order potential. Such a potential, presented in this section, is closer to real situations than

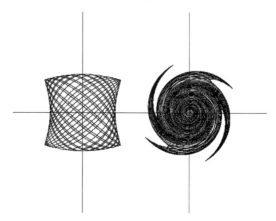

FIGURE 12.8:   Box-like orbits

the third power order potential of the Hénon-Heiles model. Box-like orbits, as well as orbits of other types, arise in this case. Figure 12.8 presents the effect of rotation applied to particles following box-like orbits (left image). The image on the right represents a four-armed spiral galaxy that results when the moving particles are rotated by using the rotation equations. As the case under discussion involves stellar bodies, the rotation angle must be selected to follow an inverse law with regards to the distance r of the particles from the centre of coordinates, as suggested in chapter 11. In this application, the following equation is used for the rotation angle:

$$\theta = 1 + \frac{1}{(r_o + r)^{3/2}}$$

The cut-off radius is equal to $r_0 = 0.1$, the simple rotation is expressed by the first term on the right, while the second term represents the influence of the attracting forces on the rotating particles.

The Hamiltonian proposed here has the form:

$$H = \frac{1}{2}(u^2 + v^2) + U(x, y)$$

where the potential $U(x, y)$ is:

$$U(x, y) = \frac{x^2}{2} + \frac{y^2}{2} + \frac{x^4}{4} - \frac{y^4}{4} - x^2 y^2$$

The parameters used in the above simulation were $x = 0$, $y = -0.25$, $u = 0.23$ and $v = 0.07$.

The equations for $\dot{u}$ and $\dot{v}$ follow from the potential function, and are

$$-\dot{u} = x + x^3 - 2xy^2$$
$$-\dot{v} = y - y^3 - 2x^2 y$$

The maximum $U(x, y)$ is achieved at $x = 1/\sqrt{5}$ and $y = \sqrt{3}/\sqrt{5}$, and it is $h = U(x, y) = 1/5$.

12.6 The Simplest Form for the Hamiltonian

The simplest form for the Hamiltonian is:

$$H = \frac{1}{2}(\dot{x}^2 + \dot{y}^2) + \frac{1}{2}(x^2 + y^2) = h$$

where the potential

$$U(x, y) = \frac{1}{2}(x^2 + y^2)$$

This gives a family of circles centred at $(x, y) = (0, 0)$ in a (x, y) diagram. The equations for $\ddot{x}$ and $\ddot{y}$ are:

$$\dot{u} = \ddot{x} = -x$$
$$\dot{v} = \ddot{y} = -y$$

It is obvious that the minimum acceptable value for the potential is 0 at $(x, y) = (0, 0)$, while there is no upper limit. The paths are ellipses and, in some cases, circles. Several paths are drawn in Figure 12.9(a) (left). The starting point is at $(x, y) = (1, 0)$, u was kept at 0, while the parameter v takes successively the values $0.1, 0.2, \ldots, 1.0$. There appear 9 ellipses and one circle for $v = 1.0$ in the diagram on the left. The diagram on the right side of the same figure illustrates the shape the orbits take after rotation with an angle:

$$\theta = \frac{0.1}{(0.001 + r)^3}$$

The outer circle of the original diagram on the left remains a circle under the rotation, but the ellipses gradually change shape, resulting in a two-armed spiral. The two-armed spiral is formed at $t = 1$, following the original formation of the circles and ellipses, as illustrated in Figure 12.9(a). If we follow the development of the curves on the right side of the figures over time, we observe that there is an anticlockwise turn of the galaxy-like object, a larger central circle, and more distinct arms. The cases for $t = 5$ and $t = 20$ are presented in Figures 12.9(b) and Figure 12.9(c).

12.7 Gravitational Attraction

The case of a small mass m attracted by a large mass M has been widely explored. Here, we assume that the large mass is located at the centre of coordinates, and it remains stable. A number of particles are attracted by the large mass from their initial position (x, y), and directed towards the mass. The particles have initial velocities $\dot{x} = u$ and $\dot{y} = v$ at position (x, y). In this case, the energy conservation equation

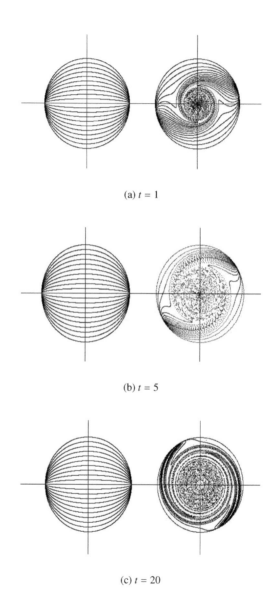

(a) $t = 1$

(b) $t = 5$

(c) $t = 20$

FIGURE 12.9: A simple Hamiltonian: orbits and rotation forms at various times

holds:

$$E = \frac{1}{2}m\left(u^2 + v^2\right) - \frac{GMm}{r}$$

or

$$\frac{u^2 + v^2}{2} - \frac{GM}{r} = \frac{E}{m} = h$$

The potential is

$$U = -\frac{GM}{r}$$

were G is the gravitational constant, and $r = \sqrt{x^2 + y^2}$. By assuming $GM = 1$, the equations of motion are:

$$\ddot{x} = \dot{u} = \frac{x}{r^3}$$

$$\ddot{y} = \dot{v} = \frac{y}{r^3}$$

In this computer experiment, we place particles with mass $m = 1$ at position $(x, y) = (1, 0)$ and initial velocities $u = 0$ and v in the interval $(0, 1)$. The $v = 1$ case corresponds to a circular path located at $(x, y) = (0, 0)$ and with radius $r = 1$. The escape velocity is $v = \sqrt{2}$. In this case, the path is a hyperbola. When $v \geq \sqrt{2}$, the particles escape to infinity. The ten paths used have velocities $v = 0.1, 0.2, \ldots, 1.0$. These paths are illustrated in Figure 12.10(a) (left). The right side of Figure 12.10(a) is the rotation image of the left side after time $t = 1$. The rotation angle follows the relation:

$$\theta = \frac{0.5}{(0.001 + r)^3}$$

The resulting shape is that of a one-armed spiral galaxy.

Another computer experiment is based on two attracting masses, located at $(x, y) = (1, 0)$ and $(x, y) = (-1, 0)$. The Hamiltonian in this case is

$$H = \frac{1}{2}(u^2 + v^2) - \frac{1}{r_1} - \frac{1}{r_2} = \frac{E}{m} = h$$

where $r_1 = \sqrt{(x - 1)^2 + y^2}$ and $r_2 = \sqrt{(x + 1)^2 + y^2}$. Excluding the diverging paths, the resulting image appears in the left part of Figure 12.10(b). In the right part of the same figure, the rotation image is illustrated after time $t = 1$. A two-armed spiral is formed. The rotation angle is of the form:

$$\theta = \frac{0.1}{(0.001 + r)^2}$$

Another two-armed spiral galaxy is formed after rotation for time $t = 1$ of the image resulting when a moving particle is attracted by two masses located at $(x, y) = (1, 0)$ and $(x, y) = (-1, 0)$, as in the previous case, but now with a slightly different rotation angle:

$$\theta = \frac{0.5}{(0.001 + r)^2}$$

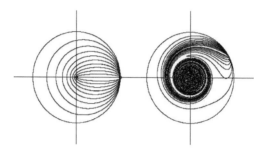

(a) A single attracting mass

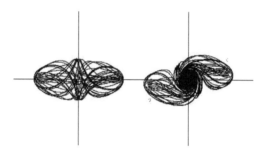

(b) Two attracting masses

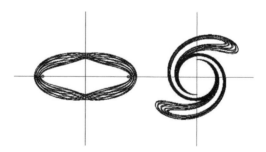

(c) Two attracting masses

FIGURE 12.10:   Orbits and rotation forms at time $t = 1$

The initial parameters were $x = 1.2$, $y = 0$, $u = 0$ and $v = 3.95$. The paths are illustrated in the left side of Figure 12.10(c), whereas in the right side the rotation object after time $t = 1$ is presented.

12.8   A Logarithmic Potential

A logarithmic form for the potential is used in this section to simulate some interesting cases in galaxy formation. The potential has the form

$$U = \frac{1}{2} \ln \left(R^2 + x^2 + c^2 y^2 \right)$$

where R is the cut-off radius and c is a parameter. Figure 12.11(a) shows a box orbit (left). The parameters are $R = 0.1$, $c = 0.8$ and the starting values are $x = 0$, $y = 1$, $u = 0.2$ and $v = 0$. The right part of Figure 12.11(a) presents the rotation of the box-like orbit with a rotation angle:

$$\theta = \frac{0.01}{(0.001 + r)^3}$$

A chaotic orbit is illustrated in Figure 12.11(b). The parameters are $R = 0.14$, $c = 1.25$ and the initial conditions are $x = 0$, $y = 0.792121$, $u = 0.1$ and $v = 0$. A loop orbit is also possible by using the above potential. The loop avoids the centre, as illustrated in the Figure 12.11(c), while the rotation image gives a two-armed spiral. The parameters are $R = 0.14$, $c = 1.25$ and the initial conditions are $x = 0$, $y = 0.2$, $u = 0.1$ and $v = 0$.

12.9   Hamiltonians with a Galactic Type Potential: The Contopoulos System

A Hamiltonian with a "galactic type" potential was first introduced by Contopoulos (1958, 1960) in his pioneering work on galaxies. The potential is based on the addition of two harmonic oscillators, along with higher order terms, to give the form:

$$U(x, y) = \frac{1}{2} \left(w_1^2 x^2 + w_2^2 y^2 \right) - exy^2$$

The resulting Hamiltonian is:

$$H = \frac{1}{2} \left(u^2 + v^2 \right) + U(x, y) = h$$

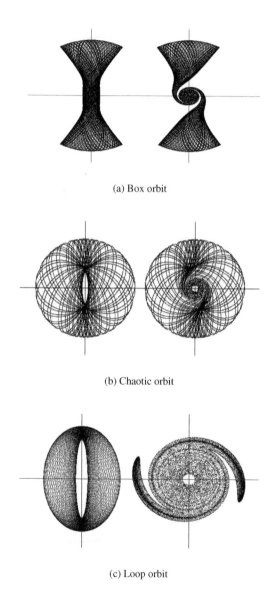

(a) Box orbit

(b) Chaotic orbit

(c) Loop orbit

FIGURE 12.11: Logarithmic potential: orbits and rotation forms at time $t = 1$

Without loss of generality this Hamiltonian can be simplified to:[2]

$$H = \frac{1}{2}\left(u^2 + v^2\right) + \frac{1}{2}\left(k^2 x^2 + y^2\right) - exy^2 = 1/2$$

where $k = w_1/w_2$ is the very important *resonance ratio*. The equations of motion are given by:

$$\dot{u} = -\frac{\partial U}{\partial x} = -k^2 x - ey^2$$

$$\dot{v} = -\frac{\partial U}{\partial y} = y - 2exy$$

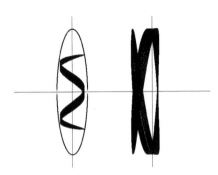

FIGURE 12.12:   The Contopoulos system: orbits at resonance ratio 4/1 (left) and 2/3 (right)

Resonant orbits characterised as 4/1 (left image) and 2/3 (right image) are illustrated in Figure 12.12. For the case on the left, the parameters are $e = 0.1$ and $k = w_1/w_2 = 4/1$, and the initial conditions are $x = 0.2$, $y = 0$, $u = 0$ and $v = 0.6$. For the case on the right, the parameters are $e = 0.1$ and $k = w_1/w_2 = 2/3$, and the initial conditions are $x = 0.1$, $y = 0$, $u = 0$ and $v = 0.9977753$.

12.10   Another Simple Hamiltonian System

A very simple Hamiltonian system that shows interesting properties has the following Hamiltonian:

$$H = \frac{1}{2}\left(u^2 + v^2\right) + \frac{1}{2}\left(x^2 + y^2\right) - \frac{1}{2}x^2 y = h$$

---

[2]Contopoulos (2002).

The potential function is:

$$U(x,y) = \frac{1}{2}\left(x^2 + y^2\right) - \frac{1}{2}x^2 y$$

The equations of motion are

$$\dot{u} = -(x - xy)$$
$$\dot{v} = -(y - x^2/2)$$

A number of equipotential curves are drawn based on the equation:

$$\frac{1}{2}\left(x^2 + y^2\right) - \frac{1}{2}x^2 y = h$$

The analysis of this equation gives the acceptable range $0 < h < 1/2$ for h. A heavy line with equation $x = \sqrt{1 + y}$ represents the $h = 1/2$ equipotential curve in Figure 12.13(a). This curve, and the line $y = 1$, define the space where the closed equipotential curves are located.

Particle orbits at high energy level for this Hamiltonian system are illustrated in Figure 12.13(b). The parameters and initial values are: $h = 1/2 - 0.001$, $x = 0.97$, $y = 0$, $u = 0$, and v is estimated from the Hamiltonian function.

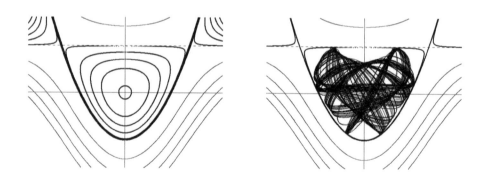

(a) Equipotential curves   (b) Orbits at high energy level

FIGURE 12.13:   Another simple Hamiltonian system

Questions and Exercises

1. Consider the set of parametric equations:

$$x = \sin(at)$$
$$y = \sin(bt)$$

   (a) Draw the curve corresponding to this system when the parameters are $a = 2$ and $b = 1$.

   (b) Determine the implicit equation for the curve in the case where $a = 2$ and $b = 1$.

   (c) Examine the curves arising for other integer values of the parameters a, b.

   (d) Consider the more general system:

$$x = \sin(at + c)$$
$$y = \sin(bt)$$

   Examine the curves arising for various integer values for a, b, and for $c = \pi/k, k \in \mathbb{N}$.

2. Consider the system:
$$x_{n+1} = x_n + a\sin(y_n)$$
$$y_{n+1} = x_n - a\sin(x_{n+1})$$

   (a) Determine the fixed points of this system, and their stability.

   (b) Compare this system to the system (12.5), and draw the characteristic graphs of this system, especially the (x_n, y_n) diagrams, for various values of the parameter a.

3. Draw the (x, y), (u, v), (x, u) and (y, v) diagrams for the discrete system (12.4) for $c = 0.1$.

4. Draw the (x, y), (u, v), (x, u) and (y, v) diagrams for the Henon-Heiles system (12.3).

Chapter 13

Odds and Ends

In this chapter we collect a number of interesting systems and examples, that we did not find an appropriate place for in the rest of this book. Topics include forced non-linear oscillators, the effect of introducing noise in three-dimensional attractors, and the Lotka-Volterra and pendulum systems. We close this chapter, and the book, with five interesting attractors.

13.1 Forced Nonlinear Oscillators

The simple two-dimensional model described by the differential equations

$$\dot{x} = -xy^2$$
$$\dot{y} = xy^2 - y \tag{13.1}$$

gives some very simple orbits in the (x, y) plane. If a *sinusoidal forcing term*

$$f(t) = a + b \cos wt$$

is added in the first equation, the system becomes

$$\dot{x} = f(t) - xy^2$$
$$\dot{y} = xy^2 - y \tag{13.2}$$

This new system can give interesting chaotic paths, as it is now three-dimensional (x, y, t). Figure 13.1 illustrates the (x, y) diagram for this system, for parameter values $a = 0.999$, $b = 0.42$ and $w = 1.75$. The resulting paths are chaotic in this case.

13.2 The Effect of Noise in Three-Dimensional Models

We discuss here the effect of adding noise to some of the models described in Chapter 6, using computer simulations. Recall that the equations of the Rössler

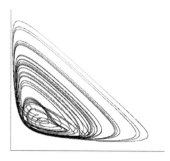

FIGURE 13.1: Adding a sinusoidal forcing term to a two-dimensional system

model are

$$\dot{x} = -y - z$$
$$\dot{y} = x - ez \qquad\qquad (13.3)$$
$$\dot{z} = f + xz - mz$$

A multiplicative noise is added in every parameter, according to the formula

$$e^* = e(1 + ku_e)$$
$$f^* = f(1 + ku_f) \qquad\qquad (13.4)$$
$$m^* = m(1 + ku_m)$$

where u_e, u_f and u_m are independent and uniformly distributed random variables in the interval $(-0.5, 0.5)$, and the noise parameter is $k = 3$.

The parameters for the simulation were set to $e = 0.2$, $f = 0.4$ and $m = 5.7$. Figure 13.2(a) illustrates the three-dimensional view of the Rössler attractor that arises. The main influence is related to the z-axis, where the maximum speed is achieved. A very interesting observation is that even strong noise does not affect the general form of the attractor much. The paths on the (x, y) plane in particular show a quite stable behaviour.

For the Lorenz model, the original equations are

$$\dot{x} = -sx + sy$$
$$\dot{y} = -xz + rz - y \qquad\qquad (13.5)$$
$$\dot{z} = xy - bz$$

The form of the Lorenz attractor when multiplicative noise is added is presented in Figure 13.2(b).

In this case, the noise is generated by a logistic process of the form:

$$f_{n+1} = 4f_n(1 - f_n)$$

This logistic process, when the chaotic parameter takes the highest value (4), may generate a uniform-like distribution, giving random numbers between zero and one.

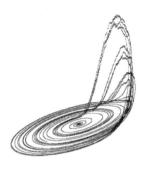

(a) Noise in the Rösler model (three-dimensional view)

(b) Noise in the Lorenz attractor ($k = 20$)

(c) Noise in the Lorenz attractor ($k = 50$)

(d) Noise in the Lorenz attractor ($k = 150$)

FIGURE 13.2: The effect of noise on three-dimensional models

The results presented here are similar to those produced when uniform noise is used instead.

The basic parameter values for the application were set to $s = 10$, $b = 8/3$ and $r = 28$, and the noise parameter was set to $k = 20$. The resulting chaotic object has a form similar to that of the Lorenz chaotic attractor. When the noise parameter k is higher ($k = 50$), the resulting chaotic image retains the Lorenz attractor shape, yet, in this case, the chaotic paths are mostly influenced by the noise term as illustrated in Figure 13.2(c). Higher values of the noise term destroy the original shape of the attractor as presented in Figure 13.2(d) where $k = 150$. At such a high noise level, the chaotic image turns out to be a stochastic one.

13.3   The Lotka-Volterra Theory for the Growth of Two Conflicting Populations

A special case of the Lotka-Volterra system was already discussed in section 6.2. We expand on it somewhat in this section.

The Lotka-Volterra system concerns the predator-prey problem, a problem frequently arising in ecology. Let N_2 be a measure of the population of a species, which preys upon a second species whose population is measured by N_1. Then the population of N_1 diminishes with a factor proportional to the product of the two populations, $N_1 N_2$, whereas the population of y increases with a factor proportional to $N_1 N_2$. The differential equations of growth or decline of the two populations are therefore given by the following set of equations:

$$\begin{aligned} \dot{N}_1 &= aN_1 - bN_1 N_2 \\ \dot{N}_2 &= -cN_2 + dN_1 N_2 \end{aligned} \tag{13.6}$$

where the parameters a, b, c, d are positive numbers. This system may be simplified by introducing the transformation:

$$\begin{aligned} N_1 &= \frac{c}{d}x \\ N_2 &= \frac{a}{b}y \end{aligned}$$

The resulting system is

$$\begin{aligned} \dot{x} &= a(x - xy) \\ \dot{y} &= -c(y - xy) \end{aligned} \tag{13.7}$$

Differentiating both equations, and eliminating y and $\dot{y}$, the following non-linear differential equation for x is obtained:

$$\ddot{x} = \frac{1}{x}\dot{x}^2 + acx - c\dot{x} + cx\dot{x} - acx^2$$

or

$$\ddot{x} = \frac{1}{x}\dot{x}^2 - c\dot{x}(1-x) + acx(1-x)$$

The equation for the phase trajectories is

$$\frac{dy}{dx} = -\frac{c(y-xy)}{a(x-xy)}$$

or

$$c\left(\dot{x} - \frac{\dot{x}}{x}\right) + a\left(\dot{y} - \frac{\dot{y}}{y}\right) = 0$$

Integrating this equation, we obtain

$$c(x - \ln x) + a(y - \ln y) = K$$

where K is the integration constant.

The limit cycles are closed curves around the equilibrium point $M = (x,y) = (1,1)$. This point is indicated by a small cycle in Figure 13.3. The parameters are $a = 1$ and $c = 3$. The initial values are $(x,y) = (2,1)$. The graph on the top of Figure 13.3 illustrates the limit cycle, whereas the (t,x) (heavy line) and the (t,y) (light line) graphs are presented in the lower part of the same figure.

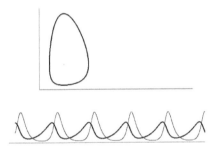

FIGURE 13.3: The Lotka-Volterra system: limit cycle (upper image) and (t,x) (heavy line) and the (t,y) (light line) graphs

The phase trajectories (x,y) for the Lotka-Volterra system are illustrated in Figure 13.4(a). The parameters are $a = 1$ and $c = 3$. Figure 13.4(b) illustrates the $(x,\dot{x})$ diagram (closed curve) and the (t,x) oscillations after a numerical solution of the non-linear second order differential equation for x presented above as a solution of the Lotka-Volterra system. The same parameter values ($a = 1$ and $c = 3$) and starting values for x and y are selected as in the previous case ($x = 2$ and $y = 1$). The resulting oscillations for x are precisely the same as in the previous case as it was expected. Similar results can be obtained after a numerical solution of the differential equation for y.

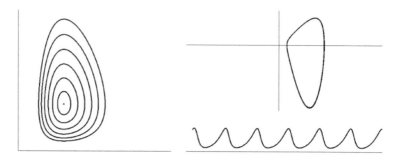

(a) (x, y) phase trajectories $(a = 1, c = 3)$ (b) $(x, \dot{x})$ diagram (closed curve) and the
     (t, x) oscillations

FIGURE 13.4: The Lotka-Volterra system

13.4 The Pendulum

Numerous applications and advances in the study of non-linear functions originate from the *pendulum differential equations*. The differential equation expressing the vibrations of the pendulum is of the form

$$\frac{d^2\theta}{dt^2} = -\frac{g}{L}\sin\theta \qquad (13.8)$$

where g is the gravitation constant, L is the length of the pendulum and θ is the angle between the position of the pendulum and the vertical direction.

When the displacement of the pendulum from the equilibrium position is small, then the angle θ is small and $\sin\theta$ may be approximated by θ. Consequently, (13.8) may be replaced with a simpler equation, expressing the behaviour for small amplitudes of the pendulum, that is

$$\frac{d^2\theta}{dt^2} = -\frac{g}{L}\theta$$

Considering that $\sin\theta = \frac{x}{L}$ and $\theta \approx \sin\theta$, the pendulum equation for θ may be replaced by a differential equation for the displacement x during time of the form

$$\frac{d^2x}{dt^2} = -\frac{g}{L^2}x$$

or

$$\frac{d^2x}{dt^2} + k^2x = 0$$

where $k = \frac{\sqrt{g}}{L}$.

This is the classical equation expressing simple harmonic oscillations. It can be solved explicitly, and the general solution has the form

$$x = A \cos kt + B \sin kt$$

or

$$x = C \cos (kt - D)$$

where A, B, C and D are constants.

The simulation of the pendulum equation is also possible by applying numerical techniques. The second order differential equation is transformed to a system of two first order differential equations of the form

$$\frac{d\theta}{dt} = f$$
$$\frac{df}{dt} = -\frac{g}{L} \sin \theta \qquad (13.9)$$

Then, the Runge-Kutta method will provide numerical solutions to the system. The position of the pendulum in a Cartesian coordinate system is located by using the expressions $x = L \sin \theta$ and $y = L \cos \theta$. Figure 13.5 illustrates the movements of the pendulum when the starting angle is $\theta = \frac{\pi}{6}$. The time variation of the horizontal displacement x is given in the lower part of the figure. The classical sinusoidal form appears.

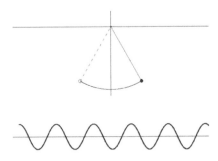

FIGURE 13.5: The pendulum and the (t, x) oscillations

Equation (13.8) can actually be integrated to give a first integral satisfying

$$\frac{d\theta}{dt} = \sqrt{\frac{2g}{L}} \sqrt{\cos \theta - \cos \omega}$$

where ω is the angle of maximum displacement of the pendulum from its equilibrium position. Figure 13.5 may also be obtained by using this formula. The final solution is given by means of elliptic integrals.

13.5   A Special Second-Order Differential Equation

Figure 13.6 illustrates the $(y, \dot{y})$ diagram of the second order differential equation

$$\ddot{y} = y^2 - 1 \tag{13.10}$$

Two equilibrium points are found. One in $(y, \dot{y}) = (1, 0)$, which is unstable, and another is $(y, \dot{y}) = (-1, 0)$, which is stable. The closed-loop curves appear around this stable fixed point. The limiting curve passes through the unstable point and the point $(y, \dot{y}) = (-2, 0)$.

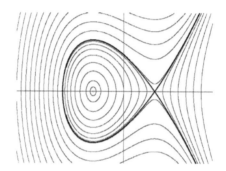

FIGURE 13.6:   The $(y, \dot{y})$ curves of a special second order differential equation system

Integration of (13.10) results in the first order differential equation

$$\dot{y}^2 = \frac{2}{3}y^3 - 2y + c$$

where c is the integration constant.

13.6   Other Patterns and Chaotic Forms

This section is a smorgasbord of chaotic attractors that are particularly interesting, but that we don't analyse in greater detail. By using the rotation formula with $b = 1$ and an equation for the rotation angle which takes into account a parallel movement of the coordinates equal to $s = 4.1$ according to the relation

$$\theta_t = c - \frac{d}{1 + (x_t - s)^2 + (y_t - s)^2}$$

a formation like the outer parts of a super-nova explosion appears (Figure 13.7). It can also be thought of as an artistic drawing presenting five dolphins in a circle. The parameters are $c = 2.8$ and $d = 13$.

FIGURE 13.7: The dolphin attractor

A graph simulating a top-down view of a ship is shown in Figure 13.8(a). It is based on a set of rotation equations of the form

$$x_{t+1} = -a - (x_t - a)\cos\theta + \frac{1}{r_t} y_t \sin\theta$$

$$y_{t+1} = -r_t x_t \sin\theta - y_t \cos\theta \qquad (13.11)$$

$$r_t = \sqrt{0.5\left(x_t^2 + \sqrt{x_t^4 + 4y_t^2}\right)}$$

The symmetry axis is located at $x = \frac{1}{2}a(\cos\theta - 1)$. The parameters are $\theta = 2$ and $a = 2.8$.

A rather startling attractor that simulates a signature is based on the following set of difference equations.

$$x_{t+1} = x_t \cos\theta_t - y_t \sin\theta_t + 1 - 0.8x_t z_t$$

$$y_{t+1} = x_t \sin\theta_t + y_t \cos\theta_t$$

$$z_{t+2} = 1.4z_{t+1} + 0.3z_t(1 - z_t) \qquad (13.12)$$

$$\theta_t = 5.5 - \frac{1}{\sqrt{x_t^2 + y_t^2 + z_t^2}}$$

A two-dimensional (x, y)-view of the simulation results appears in Figure 13.8(b).

The Ushiki model (see Ushiki, 1982) gives a form of two separated attractors (see Figure 13.9(a)) for parameter values $a = 3.64, b = 3.1, c = 0.1, d = 0.35$. The

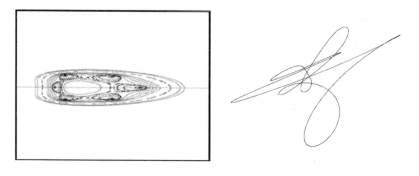

(a) A pattern of a ship   (b) The signature attractor

FIGURE 13.8:   The ship and signature attractors

equations of the model are:

$$x_{t+1} = (a - x_t - by_t)x_t$$
$$y_{t+1} = (c - y_t - dx_t)y_t$$

(13.13)

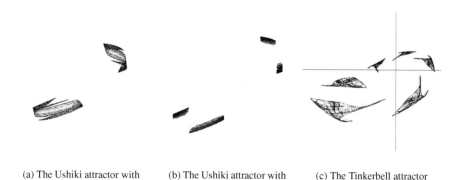

(a) The Ushiki attractor with two objects   (b) The Ushiki attractor with four objects   (c) The Tinkerbell attractor

FIGURE 13.9:   The Ushiki and Tinkerbell attractors

A four-object attractor is obtained when the parameters take the values $a = 3.6, b = 3.2, c = 0.04$ and $d = 0.4$ (Figure 13.9(b)).

The Tinkerbell model's iterations form six chaotic attractors (Figure 13.9(c)). The model is based on the following set of equations

$$x_{t+1} = x_t^2 - y_t^2 + 0.91x_t - 0.6y_t$$
$$y_{t+1} = 2x_t y_t + 2.5x_t$$

(13.14)

Questions and Exercises

1. Consider the following simple three-dimensional system of Lotka-Volterra type:

$$x' = a(x - xyz)$$
$$y' = b(-y + xyz)$$
$$z' = c(z - xyz)$$

Find the equilibrium points and the type of stability.

2. Find the line of symmetry of the ship attractor.

3. Consider the difference delay equation:

$$z_{n+2} = azn + 1 + bz_n(1 - z_n)$$

Study this equation by finding equilibrium points, exploring chaotic behaviour and drawing the related graphs and bifurcation diagrams for specific values of the parameters a and b.

4. Find the equilibrium points of the Ushiki attractor.

5. Find the equilibrium points of the Tinkerbell attractor.

Chapter 14

Milestones

It is fascinating and very instructive for any scientific field to examine the history and milestones that were the basis for the establishment and advancement of the field. Some researchers were lucky to publish and disseminate their findings to a broad audience. Others circulated their research only to a small number of friends and colleagues. And sometimes new findings came too early to be accepted by the scientific community.

During the preparation of any work an extensive bibliography is used, both directly and indirectly, by influencing the approach and direction of the research. This work was no exception. We hope the extensive bibliography, presented in the references section, will help the reader become more acquainted with the fascinating fields of chaotic modelling and simulation, and their historical development.

The table that follows presents an account of the ground-breaking steps in the development of chaotic modelling and simulation, as it is perceived in our days, through the accounts included in various books and papers. Papers written after 1980 are not included, and very few works published in the seventies appear. We have tried to emphasise those scientific works that are considered "classical," along with papers and books that described chaotic phenomena or proposed theoretical or mathematical tools that triggered the future development of the chaotic field.

Year	Name	Title of Publication
1798	T. Malthus	An essay on the principle of population
1823	M. Faraday	On a peculiar class of acoustical figures, and on certain forms assumed by groups of particles upon vibrating elastic surfaces
1825	B. Gompertz	On the nature of the function expressing of the law of human mortality
1859	P. Riess	Das Anblasen offener Rohre durch eine Flamme
—	P. L. Rijke	Notiz über eine neue Art, die in einer an beiden Enden offenen Röhre enthaltene Luft in Schwingungen zu versetzen
1883	G. Cantor	Über unendliche, lineare Punktmannigfaltigkeiten
—	Lord Rayleigh	On maintained vibrations

Year	Name	Title of Publication
—	Lord Rayleigh	On the crispations of fluid resting upon a vibrating support
1887	Lord Rayleigh	On the maintenance of vibrations by forces of double frequency, and on the propagation of waves through a medium endowed with a period structure
1889	S. Kowalevskaya	Sur le problème de la rotation d'un corps solide autour d'un point fixe
—	S. Kowalevskaya	Sur une propriété du système d'équations différentielles qui définit la rotation d'un corps solide autour d'un point fixe
1890	H. Poincaré	Mémoire sur les courbes définies par les équations différentielles I-VI, Oeuvre I
—	H. Poincaré	Sur les équations de la dynamique et le problème de trois corps
1891	Lord Rayleigh	On the problem of random vibrations, and of random flights in one, two, or three dimensions
1892	M. A. Lyapunov	Problème général de la stabilité du mouvement, (1907 translation from the 1892 Russian original)
—	H. Poincaré	Les méthodes nouvelles de la mécanique céleste
1898	J. Hadamard	Les surfaces à courbures opposées et leurs lignes géodésiques
1899	H. Poincaré	Les méthodes nouvelles de la mécanique céleste
1905	H. Poincaré	Lecons de mécanique
1908	H. Poincaré	Science et methode
1910	A. Lotka	Zur Theorie der periodischen Raktionen
1918	G. Duffing	Erzwungene Schwingungen bei veränderlicher Eigenfrequenz
—	G. Julia	Mémoire sur l'itération des fonctions rationelles
1919	F. Hausdorff	Dimension und äußeres Maß
—	P. Fatou	Sur les équations fonctionelles
1920	P. Fatou	Sur les équations fonctionelles
1923	M. H. Dulac	Sur les cycles limites
—	F. C. Ritt	Permutable rational functions
1926	P. Fatou	Sur l'itération des fonctions transcendentes entières

Year	Name	Title of Publication
1927	B. van der Pol	Forced oscillations in a circuit with non-linear resistance
—	B. van der Pol and J. van der Mark	Frequency demultiplication
1927	G. D. Birkhoff	Sur le problème restreint des trois corps
1932	G. D. Birkhoff	Sur l'existence de régions d'instabilitée dynamique
—	A. Denjoy	Sur les courbes définies par les équations différentielles à la surface du tore
1934	G. F. Gause	The struggle for existence
1935	G. D. Birkhoff	Sur le problème restreint des trois corps
—	P. O. Pedersen	Subharmonics in forced oscillations in dissipative systems
1936	A. N. Kolmogorov	Sulla teoria di Volterra della lotta per l'esistenza
1937	R. A. Fisher	The wave of advance of advantageous genes
—	A. Kolmogorov et al.	Etude de l'équation de la diffusion avec croissance de la quantitéde matière et son application à un problème biologique
1942	J. L. Doob	The Brownian movement and stochastic equations
1942	E. Hopf	Abzweigung einer periodischen Lösung von einer stationären Lösung eines Differentialsystems
1942	A. Kolmogorov	Interpolation and extrapolation of stationary series
1943	S. Chandrasekhar	Stochastic problems in physics and astronomy
1944	L. D. Landau	On the problem of turbulence
1945	M. L. Cartwright and J. E. Littlewood	On nonlinear differential equations of the second order
—	S. O. Rice	Mathematical analysis of random noise
1947	M. Kac	Random walk and the theory of Brownian motion
1948	E. Hopf	A mathematical example displaying features of turbulence
—	M. L. Cartwright	Forced oscillations in nearly sinusoidal systems
1949	C. E. Shannon and W. Weaver	The mathematical theory of information
1951	G. H. Markstein	Experimental and theoretical studies of flame-front stability

Year	Name	Title of Publication
1952	A. L. Hodgkin and A. F. Huxley	A quantitative description of membrane current and its application to conduction and excitation in nerve
—	A. L. Hodgkin and A. F. Huxley	Current carried by sodium and potassium ions through the membrane of the giant axon of loligo
1953	N. Metropolis et al.	Equations of state calculations by fast computing machine
1954	A. N. Kolmogorov	Preservation of conditionally periodic movements with small change in the Hamiltonian function
1956	W. Feller	An introduction to probability theory and its applications
1956	R. E. Kalman	Nonlinear aspects of sample data control systems
—	P. Landberg	Vibrations caused by chip formation
—	E. N. Lorenz	Empirical orthonogal functions and statistical weather prediction
—	G. Contopoulos	On the isophotes of ellipsoidal nebulae
1958	A. N. Kolmogorov	A new invariant of transitive dynamical systems
—	P. J. Myrberg	Iteration der reellen Polynome zweiten Grades
—	G. Contopoulos	On the vertical motions of stars in a galaxy
1959	N. N. Leonov	Transformations d'une droite en elle-mème
—	A. Rényi	On the dimension and entropy of probability distributions
1960	R. E. Kalman	A new approach to linear filtering and prediction problems
—	N. N. Leonov	Transformation ponctuelle d'une droite en elle-mème discontinue, linéaire par morceaux
—	N. N. Leonov	Théorie des transformations discontinues d'une droite en elle-mème
—	G. Contopoulos	A third integral of motion in a galaxy
1961	R. A. FitzHugh	Impulses and physiological states in theoretical models of nerve membrane
1962	P. J. Myrberg	Sur l'itération des polynomes réels quadratiques
—	J. Nagumo et al.	An active pulse transmission line simulating nerve axon
1963	E. N. Lorenz	Deterministic nonperiodic flow
—	B. Mandelbrot	The variation of certain speculative prices

Year	Name	Title of Publication
1964	E. N. Lorenz	The problem of deducing the climate from the governing equations
—	M. Hénon and C. Heiles	The applicability of the third integral of motion: Some numerical experiments
—	A. N. Sarkovskii	Coexistence of cycles of a continuous map of a line into itself
—	G. Contopoulos and G. Bozis	Escape of stars during the collision of two galaxies
1965	D. Coles	Transition in circular couette flow
—	A. N. Kolmogorov	Three approaches to the quantitative definition of information
—	J. R. Pasta et al.	Studies on nonlinear problems
—	L. P. Shilnikov	A case of the existence of a countable number of periodic motions
—	G. Contopoulos	Periodic and tube orbits
1967	H. Degn	Effect of bromine derivatives of malonic acid on the oscillating reaction of malonic acid, cerium ions and bromate
1967	B. Mandelbrot	The variation of some other speculative prices
—	L. P. Shilnikov	On the Poincaré-Birkhoff problem
—	L. P. Shilnikov	The existence of a denumerable set of periodic motions in four-dimensional space in an extended neighbourhood of a saddle-focus
1968	M. Kuczma	Functional equations in a single variable
—	I. Prigogine and R. Lefever	Symmetry breaking instabilities in dissipative systems. II
1969	R. FitzHugh	Mathematical models of excitation and propagation in nerve and muscle
—	M. Hénon	Numerical study of quadratic area-preserving mappings
–	E. N. Lorenz	Atmospheric predictability as revealed by naturally occurring analogies
—	I. Prigogine et al.	Symmetry breaking instabilities in biological systems
1976	O. E. Rössler	Chaotic behavior in simple reaction systems

Year	Name	Title of Publication
—	O. E. Rössler	Chemical turbulence: Chaos in a simple reaction-diffusion system
—	O. E. Rössler	Different types of chaos in two simple differential equations
1978	M. J. Feigenbaum	Quantitative universality for a class of nonlinear transformations
1979	M. J. Feigenbaum	The universal metric properties of nonlinear transformations
—	O. E. Rössler	An equation for hyperchaos
—	O. E. Rössler	Continuous chaos — four prototype equations
—	K. Ikeda	Multiple-valued stationary state and its instability of the transmitted light by a ring cavity system
1980	K. Ikeda et al.	Optical turbulence: Chaotic behaviour of transmitted light from a ring cavity

References

Acheson, D. (1997). *From calculus to chaos*. Oxford: Oxford University Press.

Adachi, S., M. Toda, and K. Ikeda (1988a). Potential for mixing in quantum chaos. *Phys. Rev. Lett. 61*, 635.

Adachi, S., M. Toda, and K. Ikeda (1988b). Quantum-classical correspondence in many-dimensional quantum chaos. *Phys. Rev. Lett. 61*, 659.

Adler, M. and P. van Moerbeke (1994). The Kowalevski and Hénon-Heiles motions as Manakov geodesic flows on SO(4). *Comm. Math. Phys. 113*, 649.

Alligood, K. T. and T. Sauer (1988). Rotation numbers of periodic orbits in the Hénon's map. *Commun. Math. Phys. 120*, 105.

Alligood, K. T., T. D. Sauer, and J. A. Yorke (1997). *Chaos; An introduction to dynamical systems*. Berlin: Springer-Verlag.

Almirantis, Y. and M. Kaufman (1992). Numerical study of travelling waves in a reaction-diffusion system: response to a spatiotemporal forcing. *Int. J. of Bifurcation and Chaos 2*(1), 51–60.

Almirantis, Y. and M. Kaufman (1995). Chiral selection of rotating waves in a reaction-diffusion system: The effect of a circularly polarized electromagnetic field. *Int. J. of Bifurcation and Chaos 5*(2), 507–518.

Almirantis, Y. and G. Nicolis (1987). Morphogenesis in an asymmetric medium. *Bull. Math. Biol. 47*, 519–530.

Alsing, P. M., A. Gavrielides, and V. Kovanis (1994a). History-dependent control of unstable periodic orbits. *Phys. Rev. E 50*, 1968.

Alsing, P. M., A. Gavrielides, and V. Kovanis (1994b). Using neural networks for controlling chaos. *Phys. Rev. E 49*(2), 1225–1231.

Androulakakis, S. P., B. Greenspan, and T. T. H. H. Qammar (1991). Practical considerations on the calculation of the uncertainty exponent and the fractal dimension of basin boundaries. *Int. J. of Bifurcation and Chaos 1*(2), 327–333.

Antoniou, I., V. Basios, and F. Bosco (1996). Probabilistic control of chaos: The beta-adic Rényi map under control. *Int. J. of Bifurcation and Chaos 6*(8), 1563–1573.

Aref, H. (1983). Integrable, chaotic, and turbulent vortex motion in two-dimensional flows. *Ann. Rev. Fluid Mech. 15*, 345.

Aref, H. (1984). Stirring by chaotic advection. *J. Fluid mech. 143*, 1.

Aref, H. and S. Balachandar (1986). Chaotic advection in a Stokes flow. *Phys. Fluids 29*, 3515–3521.

Argyris, J. and I. Andreadis (1998). On linearisable noisy systems. *Chaos Solitons Fractals 9*(6), 895–899.

Argyris, J. and I. Andreadis (2000). On the influence of noise on the coexistence of chaotic attractors. *Chaos Solitons Fractals 11*(6), 941–946.

Argyris, J. and C. Ciubotariu (1999). A new physical effect modeled by an ikeda map depending on a monotonically time-varying parameter. *Int. J. of Bifurcation and Chaos 9*(6), 1111–1120.

Argyris, J., B. V., N. Ovakimyan, and M. Minasyan (1996). Chaotic vibrations of a nonlinear viscoelastic beam. *Chaos Solitons Fractals 7*(2), 151–163.

Arnéodo, A., F. Argoul, J. Elezgaray, and P. Richetti (1993). Homoclinic chaos in chemical systems. *Physica D 62*, 134–169.

Arneodo, A., P. Coullet, and C. Tresser (1979). A renormalization group with periodic behavior. *Phys. Lett. A 70*, 74.

Arneodo, A., P. Coullet, and C. Tresser (1980). Occurrence of strange attractors in three-dimensional Volterra equations. *Phys. Lett. A 79*, 59.

Arneodo, A., P. Coullet, and C. Tresser (1981a). A possible new mechanism for the onset of turbulence. *Phys. Lett. A 81*, 197.

Arneodo, A., P. Coullet, and C. Tresser (1981b). Possible new strange attractors with spiral structure. *Commun. Math. Phys. 79*, 573–579.

Arnold, L. (1990). Stochastic differential equations as dynamical systems. In M. K. et al. (Ed.), *Proceedings MTNS-89 Amsterdam*, Volume I, Boston, pp. 489–495. Birkhäuser.

Arnold, L. (1998). *Random Dynamical Systems*. Berlin: Springer-Verlag.

Arnold, V. I. (1983). *Geometrical methods in the theory of ordinary differential equations*. Berlin: Springer-Verlag.

Aronson, D. G. (1980). Density dependent reaction-diffusion system. In *Dynamics and modelling of reactive systems*, pp. 161–176. New York: Academic Press.

Aronson, J. (1990). CHAOS: A SUN-based program for analyzing chaotic systems. *Computers in Physics 4*(4), 408–417.

Arrowsmith, D. K. and C. M. Place (1990). *An introduction to dynamic systems*. Cambridge: Cambridge University Press.

Aston, P. J. (1998). A chaotic Hopf bifurcation in coupled maps. *Physica D 118*(3-4), 199–220.

Aston, P. J. (1999). Bifurcations of the horizontally forced spherical pendulum. *Comput. Method. Appl. Mech. Eng. 170*(3-4), 343–353.

Auerbach, D., C. Grebogi, E. Ott, and J. Yorke (1992). Controlling chaos in high dimensional systems. *Phys. Rev. Lett. 69*, 3479–3482.

Awrejcewicz, J. (1989). *Bifurcation and chaos in simple dynamical systems.* Singapore: World Scientific.

Bahar, S. (1996a). Further studies of bifurcations and chaotic orbits generated by iterated function systems. *Chaos Solitons Fractals 7*(1), 41–47.

Bahar, S. (1996b). Patterns of bifurcation in iterated function systems. *Chaos Solitons Fractals 7*(2), 205–210.

Bahar, S. (1997). Orbits embedded in IFS attractors. *Int. J. of Bifurcation and Chaos 7*(3), 741–749.

Balakrishnan, V., C. Nicolis, and G. Nicolis (1995). Extreme value distributions in chaotic dynamics. *J. Stat. Phys. 80*(1-2), 307–336.

Balakrishnan, V., G. Nicolis, and C. Nicolis (1997). Recurrence time statistics in chaotic dynamics. I. discrete time maps. *J. Stat. Phys. 86*(1-2), 191–212.

Balakrishnan, V., G. Nicolis, and C. Nicolis (2000). Recurrence time statistics in deterministic and stochastic dynamical systems in continuous time: A comparison. *Phys. Rev. E 61*(3), 2490–2499.

Basios, V., T. Bountis, and G. Nicolis (1999). Controlling the onset of homoclinic chaos due to parametric noise. *Phys. Lett. A 251*(4), 250–258.

Bass, F. (1969). A new product growth model for consumer durables. *Management Science 15*(217–231).

Belyakova, G. V. and L. A. Belyakov (1997). On bifurcations of periodic orbits in the van der Pol-Duffing equation. *J. of Bifurcation and Chaos 7*(2), 459–462.

Bergé, P., M. Dubois, P. Manneville, and Y. Pomeau (1980). Intermittency in Rayleigh-Bénard convection. *J. Phys. Lett. 41*, 341.

Bergé, P. and Y. Pomeau (1980). La turbulence. *La Recherche 11*, 422.

Bier, M. and T. C. Bountis (1984). Remerging Feigenbaum trees in dynamical systems. *Phys. Lett. A 104*, 29.

Biktashev, V. N. (1989). Evolution of twist of an autowave vortex. *Physica D 36*, 167–172.

Birkhoff, G. and G.-C. Rota (1989). *Ordinary differential equations.* New York, N.Y.: Wiley.

Birkhoff, G. D. (1927). On the periodic motions of dynamical systems. *Acta Math. (reprinted in MacKay and Meiss 1987) 50*, 359.

Birkhoff, G. D. (1932). Sur l'existence de régions d'instabilitée dynamique. *Ann. Inst. H. Poincaré 2*, 369.

Birkhoff, G. D. (1935). Sur le problème restreint des trois corps. *Ann. Scuola Norm. Sup. Pisa 4*, 267.

Bolotin, V. V., A. A. Grishko, A. N. Kounadis, C. Gantes, and J. B. Roberts (1998). Influence of initial conditions on the postcritical behavior of a nonlinear aeroelastic system. *Nonlinear Dyn. 15*(1), 63–81.

Borland, L. (1996). Simultaneous modeling of nonlinear determistic and stochastic dynamics. *Physica D 99*(2-3), 175–190.

Borland, L. (1998). Microscopic dynamics of the nonlinear Fokker-Planck equation: A phenomenological model. *Phys. Rev. E 57*(6), 6634–6642.

Borland, L. and H. Haken (1992). Unbiased determination of forces causing observed processes. The case of additive and weak multiplicative noise. *Z. Phys. B - Condens. Matter 81*, 95.

Borland, L. and H. Haken (1993a). Learning the dynamics of two-dimensional stochastic Markov processes. *Open Syst. and Inf. Dyn. 1*(3), 311.

Borland, L. and H. Haken (1993b). On the constraints necessary for macroscopic prediction of stochastic time-dependent processes. *ROMP 33*, 35.

Boudourides, M. A. and N. A. Fotiades (2000). Piecewise linear interval maps both expanding and contracting. *Dyn. Stab. Syst. 15*(4), 343–351.

Bountis, T. (1992). *Chaotic dynamics. Theory and practice*. New York: Plenum Press.

Bountis, T., L. Drossos, and I. C. Percival (1991a). Nonintegrable systems with algebraic singularities in complex time. *J. Phys. 24*, 3217.

Bountis, T., L. Drossos, and I. C. Percival (1991b). On nonintegrable systems with square root singularities in complex time. *Phys. Lett. A 159*, 1.

Bountis, T. and R. H. Helleman (1981). On the stability of periodic orbits of two-dimensional mappings. *J. Math. Phys. 22*, 1867.

Bountis, T., L. Karakatsanis, G. Papaioannou, and G. Pavlos (1993). Determinism and noise in surface temperature time series. *Ann. Geophys. 11*, 947–959.

Bountis, T., V. Papageorgiou, and M. Bier (1987). On the singularity analysis of intersecting separatrices in near-integrable dynamical systems. *Physica D 24*, 292.

Bountis, T., H. Segur, and F. Vivaldi (1982). Integrable Hamiltonian systems and the Painlevé property. *Phys. Rev. A 25*, 1257.

Bountis, T. C. (1981). Period doubling bifurcations and universality in conservative systems. *Physica D 3*, 577–589.

Boyarsky, A. (1986). A functional equation for a segment of the Hénon map unstable manifold. *Physica D 21*, 415.

Braun, T. and I. A. Heisler (2000). A comparative investigation of controlling chaos in a Rössler system. *Physica A 283*(1-2), 136–139.

Briggs, K. M. (1989). How to calculate the Feigenbaum constants on your PC. *Aust. Math Soc. Gazette 16*, 89–92.

Brock, W. A. and C. H. Hommes (1997, August). Heterogeneous beliefs and routes to chaos in a simple asset pricing model. *Journal of Economic Dynamics and Control 22*(8-9), 1235–1274.

Brock, W. A., J. Lakonishok, and B. LeBaron (1992). Simple technical trading rules and the stochastic properties of stock returns. *J. Finance 47*, 1731–1764.

Brock, W. A. and B. LeBaron (1993). Using structural modelling in building statistical models of volatility and volume of stock market returns. Technical report, Univ. of Wisconsin, Madison, Madison, Wisconsin.

Brock, W. A. and A. G. Malliaris (1989). *Differential equations, stability and chaos in dynamic economics*. Amsterdam: North Holland.

Budinsky, N. and T. Bountis (1983). Stability of nonlinear modes and chaotic properties of 1D Fermi-Pasta-Ulam lattices. *Physica D 8*, 445–452.

Bunner, M. J. (1999). The control of high-dimensional chaos in time-delay systems to an arbitrary goal dynamics. *Chaos 9*(1), 233–237.

Busse, H. (1969). A spatial periodic homogeneous chemical reaction. *J. Phys. Chem. 73*, 750.

Cabrera, J. L. and F. J. de la Rubia (1996). Analysis of the behavior of a random nonlinear delay discrete equation. *Int. J. of Bifurcation and Chaos 6*(9), 1683–1690.

Calogero, F. (1971). Solution of the 1-D N-body problem with quadratic and/or inversely quartic pair potentials. *J. Math. Phys. 12*, 419.

Campanino, M., H. Epstein, and D. Ruelle (1982). On Feigenbaum's functional equation. *Topology 21*, 125–129.

Caranicolas, N. and C. Vizikis (1987). Chaos in a quartic dynamical model (celestial mechanics). *Celest. Mech. 40*, 35.

Caratheodory, C. (1982). *Calculus of variations and partial differential equations of the first order*. New York: Chelsea.

Carroll, T. L. (1994). Synchronization of chaos. *Ciencia Hoje 18*, 26.

Carroll, T. L. (2002). Noise-resistant chaotic maps. *Chaos 12*(2), 275–278.

Carroll, T. L. and L. M. Pecora (1999). Using multiple attractor chaotic systems for communication. *Chaos 9*(2), 445–451.

Cartwright, M. L. (1948). Forced oscillations in nearly sinusoidal systems. *J. Inst. Elec. Eng. 95*, 88.

Cartwright, M. L. and J. E. Littlewood (1945). On nonlinear differential equations of the second order. *J. London Math. Soc. 20*, 180–189.

Casdagli, M. (1989). Nonlinear prediction of chaotic time series. *Physica D 35*, 335–356.

Casdagli, M. (1991). Chaos and deterministic "versus" stochastic non-linear modelling. *J. Roy. Statist. Soc. Ser. B 54*(2), 303–328.

Casdagli, M. C. and A. S. Weigend (1993). Exploring the continuum between deterministic and stochastic modeling. In A. S. Weigend and N. A. Gershenfeld (Eds.), *Time series prediction: Forecasting the future and understanding the past*, pp. 347–366. Reading, MA: Addison-Wesley.

Catsigeras, E. (1996). Cascades of period doubling bifurcations in *n* dimensions. *Nonlinearity 9*, 1061–1070.

Catsigeras, E. and H. Enrich (1999). Persistence of the Feigenbaum attractor in one-parameter families. *Commun. Math. Phys. 207*(3), 621–640.

Caurier, E. and B. Grammaticos (1986). Quantum chaos with nonergodic Hamiltonians. *Europhys. Lett. 2*, 417.

Caurier, E. and B. Grammaticos (1989). Extreme level repulsion for chaotic quantum Hamiltonians. *Phys. Lett. A 136*, 387–390.

Celikovsky, S. and G. R. Chen (2002). On a generalized Lorenz canonical form of chaotic systems. *Int. J. Bifurcation Chaos 12*(8), 1789–1812.

Celka, P. (1996). A simple way to compute the existence region of 1D chaotic attractors in 2D-maps. *Physica D 90*(3), 235–241.

Celka, P. (1997). Delay-differential equation versus 1D-map: Application to chaos control. *Physica D 104*(127-147).

Chandrasekhar, S. (1943). Stochastic problems in physics and astronomy. *Mod. Phys. 15*(1), 1–89.

Chirikov, B. (1996). Natural laws and human prediction. In P. Weingartner and G. Schurz (Eds.), *Law and prediction in the light of chaos research*, Volume LNP 473, pp. 10–34. Berlin: Springer-Verlag.

Chirikov, B. V. (1979). A universal instability of many-dimensional oscillator systems. *Phys. Rep. 52*, 263.

Chirikov, B. V. (1983). Chaotic dynamics in Hamiltonian systems with divided phase space. In *Dynamical systems and chaos*, Volume 179 of *Lecture notes in physics*, pp. 29–46.

Chorafas, D. (1994). *Chaos theory in the financial markets. Applying fractals. Fuzzy logic. Genetic algorithms.* Probus Publishing Co.

Chua, L. O. (1980). Dynamic nonlinear networks: State of the art. *IEEE Trans. Circuit Syst. 27*(11), 1059–1087.

Chua, L. O. (1992). The genesis of Chua's circuit. *Archiv für Elektronik & Ü.-technik 46*, 250–257.

Chua, L. O. (1993a). Global unfolding of Chua's circuit. *IEICE Trans. Fundamentals E 76-A*, 704–734.

Chua, L. O. (1993b). A zoo of strange attractors from the canonical Chua's circuit. *J. Circuit, Systems, and Computers 2*.

Chua, L. O., M. Itoh, L. Kocarev, and K. Eckert (1993). Chaos synchronization in Chua's circuit. *J. Circuits, Systems and Computers 3*(1), 93–108.

Chung, J. S. and M. M. Bernitsas (1992). Dynamics of two-line ship towing/mooring systems: bifurcations, singularities, instability boundaries, chaos. *J. Ship Res. 36*(2).

Cladis, P. E. and P. Palffy-Muhoray (1994). *Spatio-Temporal Patterns in Nonequilibrium Complex Systems (SFI Studies in the Sciences of Complexity)*. Reading, MA: Addison-Wesley.

Clerc, M., P. Coullet, and E. Tirapegui (1999). Lorenz bifurcation: Instabilities in quasireversible systems. *Phys. Rev. Lett. 83*(19), 3820–3823.

Coleman, M. J. and A. Ruina (1998). An uncontrolled walking toy that cannot stand still. *Phys. Rev. Lett. 80*(16), 3658–3661.

Coleman, S. (1993). Cycles and chaos in political party voting. *Journal of Mathematical Sociology 18*, 47–64.

Coleman, S. (1995). Dynamics in the fragmentation of political party systems. *Qualilty and Quantity 29*, 141–155.

Coles, D. (1965). Transition in circular couette flow. *J. of Fluid Mech. 21*, 385–425.

Collet, P. and J.-P. Eckmann (1980a). *Iterated maps on the interval as dynamical system*. Basel: Birkhäuser.

Collet, P. and J.-P. Eckmann (1980b). On the abundance of chaotic behavior in one dimension. In R. Helleman (Ed.), *Nonlinear Dynamics*, Volume 357 of *Ann. N. Y. Acad. Sci.*, pp. 337–342. New York: The New York Academy of Sciences.

Collet, P. and J.-P. Eckmann (1983). Positive Liapunov exponents and absolute continuity for maps of the interval. *Ergodic Theory & Dynamical Systems 3*, 13–46.

Collet, P., J.-P. Eckmann, and H. Koch (1981a). On universality for area-preserving maps of the plane. *Physica D 3*, 457.

Collet, P., J.-P. Eckmann, and H. Koch (1981b). Period doubling bifurcations for famlies of maps on $\mathbb{R}^n$. *J. Stat. Phys. 25*, 1.

Collet, P., J.-P. Eckmann, and O. E. Lanford (1980). Universal properties of maps on an interval. *Commun. Math. Phys. 76*, 211–254.

Collet, P., J.-P. Eckmann, and L. Thomas (1981). A note on the power spectrum of the iterates of Feigenbaum's function. *Commun. Math. Phys. 81*, 261–265.

Combes, F., F. Debbasch, D. Friedli, and D. Pfenniger (1990). Box and peanut shapes generated by stellar bars. *Astron. Astrophys. 233*, 82.

Contopoulos, G. (1956). On the isophotes of ellipsoidal nebulae. *Z. Astrophysic 39*, 126.

Contopoulos, G. (1958). On the vertical motions of stars in a galaxy. *Stockholm Ann. 20 No5.*

Contopoulos, G. (1960). A third integral of motion in a galaxy. *Z. Astrophysic 49*, 273–291.

Contopoulos, G. (1965). Periodic and tube orbits. *Astron. J. 70*, 526.

Contopoulos, G. (1981). Do successive bifurcations in Hamiltonian systems have the same universal ratio? *Lett. Nuovo Cimento 30*, 498.

Contopoulos, G. (2001). The development of nonlinear dynamics in astronomy. *Found. Phys. 31*(1), 89–114.

Contopoulos, G. (2002). *Order and chaos in dynamical astronomy*. Berlin: Springer-Verlag.

Contopoulos, G. and G. Bozis (1964). Escape of stars during the collision of two galaxies. *Astrophys. J. 139*, 1239.

Contopoulos, G., M. Hénon, and D. Lynden-Bell (Eds.) (1973). *Dynamic structure and evolution of stellar systems*. Saas.

Contopoulos, G. and C. Polymilis (1987). Approximations of the 3-particle Toda lattice. *Physica D 24*, 328.

Coullet, P. and F. Plaza (1995). Excitable spiral waves in nematic liquid crystals. *Int. J. of Bifurcation and Chaos 4*(5), 1173–1182.

Coullet, P., C. Tresser, and A. Arneodo (1979). Transition to stochasticity for a class of forced oscillators. *Phys. Lett. A 72*, 268.

Creedy, J. (1994). *Chaos and non-linear models in economics: theory and applications*. Aldershot: Elgar.

Crutchfield, J. P. (1983). *Noisy chaos*. Ph.D. dissertation, University of California,

Santa Cruz.

Crutchfield, J. P. and B. A. Huberman (1980). Fluctuations and the onset of chaos. *Phys. Lett. A 77*, 407–410.

Curry, J. H. (1981). On computing the entropy of the Hénon attractor. *J. Stat. Phys. 26*, 683.

Cvitanović, P. (1988). Invariant measures of strange sets in terms of cycles. *Phys. Rev. Lett. 61*, 2729–2732.

Cvitanović, P. (1989). *Universality in chaos*. Bristol: Adam Hilger.

Cvitanović, P. (1991). Periodic orbits as the skeleton of classical and quantum chaos. *Physica D 51*, 138–151.

Cvitanović, P. and B. Eckhardt (1989). Periodic orbit quantization of chaotic systems. *Phys. Rev. Lett. 63*, 823–826.

Cvitanović, P., P. Gaspard, and T. Schreiber (1992). Investigation of the Lorentz gas in terms of periodic orbits. *Chaos 2*, 85.

Cvitanović, P., G. H. Gunaratne, and M. J. Vinson (1990). On the mode-locking universality for critical circle maps. *Nonlinearity 3*, 873–885.

Cvitanović, P., K. Hansen, J. Rof, and G. Vattay (1998). Beyond the periodic orbit theory. *Nonlinearity 11*, 1209–1232.

Cvitanović, P. and J. Myrheim (1983). Universality for period n-tuplings in complex mappings. *Phys. Lett. A 94*, 329–333.

Cybenko, G. (1989). Approximation by superpositions of a sigmoidal function. *Math. Control, Signals Syst 2*, 303–314.

Daido, H. (1980). Analytic conditions for the appearance of homoclinic and heteroclinic points of a 2-dimensional mapping: the case of the Hénon mapping (1831). *Prog. Theor. Phys. 63*, 1190.

Daido, H. (1981). Theory of the period-doublings of 1-D mappings based on the parameter dependence. *Phys. Lett. A 83*, 246.

Daido, H. (1997). Strange waves in coupled-oscillator arrays: Mapping approach. *Phys. Rev. Lett. 78(9)*, 1583–1686.

Davies, M. (1994). Noise reduction schemes for chaotic time series. *Physica D 79*, 174–192.

Davies, M. E. (1992). An order N noise reduction algorithm with quadratic convergence. Preprint.

Davis, H. T. (1960). *Introduction to nonlinear differential and integral equations*. New York: Dover.

Davis, S. H. (1987). Coupled Lorenz oscillators. *Physica D 24*, 226.

Day, R. H. (1994). *Complex economic dynamics*. Cambridge, MA: MIT Press.

Day, R. H. and M. Zhang (1996). Classical economic growth theory: a global bifurcation analysis. *Chaos Solitons Fractals 7*(12), 1969–1988.

de la Llave, R. and S. Tompaidis (1995). On the singularity structure of invariant curves of sympletic mappings. *Chaos 5*(1), 227–237.

de Olivera, C. R. and C. P. Malta (1987). Bifurcations in a class of time-delay equations. *Phys. Rev. A 36*, 3997–4001.

Degn, H. (1967). Effect of bromine derivatives of malonic acid on the oscillating reaction of malonic acid, cerium ions and bromate. *Nature 213*, 589–590.

Dellnitz, M., M. Field, M. Golubitsky, J. Ma, and A. Hoohmann (1995). Cycling chaos. *Int. J. of Bifurcation and Chaos 5*(4), 1243–1247.

Dellnitz, M., M. Golubitsky, A. Hohmann, and I. Stewart (1995). Spirals in scalar reaction-diffusion equations. *Int. J. of Bifurcation and Chaos 5*(6), 1487–1501.

Dendrinos, D. S. (1994). Traffic-flow dynamics: a search for chaos. *Chaos Solitons Fractals 4*, 605–617.

Dendrinos, D. S. and M. Sonis (1990). *Chaos and social-spatial dynamics*. New York: Springer-Verlag- Verlag.

Denjoy, A. (1932). Sur les courbes définies par les équations différentielles à la surface du tore. *J. Math. Pures Appl. 11*, 333–375.

Derrida, B. and Y. Pomeau (1980). Feigenbaum's ratios of two-dimensional area preserving maps. *Phys. Lett. A 80*, 217–219.

Devaney, R. (1976). Reversible diffeomorphisms and flows. *Trans. Amer. Math. Soc. 218*, 89–113.

Devaney, R. L. (1988). Chaotic bursts in nonlinear dynamical systems. *Science 235*, 342–345.

Devaney, R. L. (1989). *An introduction to chaotic dynamical systems*. Reading, MA: Addison-Wesley.

Devaney, R. L. (1991). e^z: Dynamics and bifurcations. *Int. J. of Bifurcation and Chaos 1*(2), 287–308.

Devaney, R. L. and M. Durkin (1991). The exploding exponential and other chaotic bursts in complex dynamics. *Amer. Math. Monthly 98*, 217–233.

Devaney, R. L. and X. Jarque (1997). Misiurewicz points for complex exponentials. *Int. J. of Bifurcation and Chaos 7*(7), 1599–1615.

Dewar, R. L. and A. B. Khorev (1995). Rational quadratic-flux minimizing circles for area-preserving twist maps. *Physica D 85*(1-2), 66–78.

Diakonos, F. K., D. Pingel, and P. Schmelcher (1999). A stochastic approach to the construction of one-dimensional chaotic maps with prescribed statistical properties. *Phys. Lett. A 264*(2-3), 162–170.

Diakonos, F. K., P. Schmelcher, and O. Biham (1998). Systematic computation of the least unstable periodic orbits in chaotic attractors. *Phys. Rev. Lett. 81*(20), 4349–4352.

Diamond, P. (1994). Chaos in iterated fuzzy systems. *J. Math. Anal. Appl. 184*, 472–484.

Dimarogonas, A. D. (1996). Vibration of cracked structures: a state of the art review. *Engineering Fracture Mechanics 55*(5), 831–857.

Ding, M., E. Ott, and C. Grebogi (1994). Crisis control: preventing chaos-induced capsizing of a ship. *Phys. Rev. E 50*(5), 4228–4230.

Ditto, W. L. (1996). Applications of chaos in biology and medicine. *AIP Conf. Proc. 376*, 175–201.

Doering, C. R. and J. D. Gibbon (1998). On the shape and dimension of the Lorenz attractor (vol 10, pg 255, 1995). *Dynam. Stabil. Syst. 13*(3), 299–301.

Doherty, M. F. and J. M. Ottino (1988). Chaos in deterministic systems: strange attractors, turbulence, and applications in chemical engineering. *Chem. Eng. Sci. 43*, 139.

Dokoumetzidis, A., A. Iliadis, and P. Macheras (2001). Nonlinear dynamics and chaos theory: concepts and applications relevant to pharmacodynamics. *Pharm. Res. 18*(4), 415–426.

Domokos, G. and P. J. Holmes (1993). Euler's problem and Euler's method, or the discrete charm of buckling. *J. Nonlinear Sci. 3*, 109–151.

Doob, J. L. (1942). The Brownian movement and stochastic equations. *Ann. of Math. 43*(2), 351–369.

Dormayer, P. (1994). Examples for smooth bifurcation for $x(t) = -alphaf(x(t-1))$. *Appl. Anal. 55*(1-2), 25–40.

Drossos, L. and T. Bountis (1992). On the convergence of series solutions of nonintegrable systems with algebraic singularities. In T. Bountis (Ed.), *Chaotic dynamics and practice*. London: Plenum.

Drossos, L., O. Ragos, M. N. Vrahatis, and T. Bountis (1996). Method for computing long periodic orbits of dynamical systems. *Phys. Rev. E 53*(1), 1206–2111.

Duffing, G. (1918). *Erzwungene Schwingungen bei veränderlicher Eigenfrequenz.* Braunschweig: Vieweg & Sohn.

Dulac, M. H. (1923). Sur les cycles limites. *Bull. Soc. Math. Anal. 51*, 45–188.

Eckmann, J.-P. (1981). Roads to turbulence in dissipative dynamical systems. *Rev. Mod. Phys. 53*, 643–654.

Eckmann, J.-P., H. Koch, and P. Wittwer (1984). A computer-assissted proof of universality for area-preserving maps. *Mem. Am. Math. Soc. 47*, 1–122.

Eckmann, J.-P. and D. Ruelle (1985). Ergodic theory of chaos and strange attractors. *Rev. Mod. Phys. 57*, 617–656.

Eckmann, J.-P. and P. Wittwer (1987). A complete proof of the Feigenbaum conjectures. *J. Stat. Phys. 46*, 455–477.

El-Rifai, E. A. and E. Ahmed (1995). Knotted periodic orbits in Rössler's equations. *J. Math. Phys. 36*(2), 773–777.

Elsner, J. B. and A. A. Tsonis (1992). Nonlinear predicting, chaos and noise. *Bull. Amer. Meteor. Soc. 73*, 49–60.

Endler, A. and J. A. C. Gallas (2001). Period four stability and multistability domains for the Hénon map. *Physica A 295*(1-2), 285–290.

Epstein, H. (1986). New proofs of the existence of the Feigenbaum functions. *Comm. Math. Phys. 106*, 395–426.

Epstein, H. R. (1983). Oscillations and chaos in chemical systems. *Physica D 7*, 47.

Fang, H. P. (1994). Studying the Lorenz equations with one-dimensional maps from successive local maxima in z. *Z. Phys. B 96*(4), 547–552.

Farady, M. (1831). On a peculiar class of acoustical figures, and on certain forms assumed by groups of particles upon vibrating elastic surfaces. *Phil. Trans. Roy. Soc. London 121*, 299–340.

Fatou, P. (1919). Sur les équations fonctionelles. *Bull. Soc. Math. France 47*, 161–271.

Fatou, P. (1920). Sur les équations fonctionelles. *Bull. Soc. Math. France 48*, 33–94, 208–314.

Fatou, P. (1926). Sur l'itération des fonctions transcendentes entières. *Acta Math. 47*, 337–370.

Feichtinger, G. (1992a). Hopf bifurcation in an advertising diffusion model. *J. Econ. Behav. Organ. 17*, 401–411.

Feichtinger, G. (1992b). Limit cycles in dynamic economic systems. *Ann. Operation Res. 37*, 313–344.

Feichtinger, G., L. L. Ghezzi, and C. Piccardi (1995). Chaotic behavior in an advertising diffusion model. *Int. J. of Bifurcation and Chaos 5*(1), 255–263.

Feigenbaum, M. J. (1978). Quantitative universality for a class of nonlinear transformations. *J. Stat. Phys. 19*, 25–52.

Feigenbaum, M. J. (1979). The universal metric properties of nonlinear transformations. *J. Stat. Phys. 21*, 669–706.

Feigenbaum, M. J. (1980a). *The metric universal properties of period doubling bifurcations and the spectrum for a route to turbulence.*, Volume 357 of *Ann. N. Y. Acad. Sci.* New York: The New York Academy of Sciences.

Feigenbaum, M. J. (1980b). The onset spectrum of turbulence. *Phys. Lett. A 74*, 375–378.

Feigenbaum, M. J. (1980c). The transition to aperiodic behaviour in turbulent systems. *Commun. Math. Phys. 77*, 65–86.

Feigenbaum, M. J. (1980d). Universal behaviour in nonlinear systems. *Los Alamos Science 1*, 4–27.

Feigenbaum, M. J. (1983). Universal behavior in nonlinear systems. *Physica D 7*, 16–39.

Feigenbaum, M. J. (1987a). Scaling spectra and return times of dynamical systems. *J. Stat. Phys. 46*, 925–932.

Feigenbaum, M. J. (1987b). Some characterizations of strange sets. *J. Stat. Phys. 46*, 919–924.

Feigenbaum, M. J. (1988). Presentation functions, fixed points, and a theory of scaling function dynamics. *J. Stat. Phys. 52*, 527–569.

Feigenbaum, M. J. (1990). Presentation functions and scaling function theory for circle maps. In D. K. Campbell (Ed.), *Chaos — Soviet American perspectives on nonlinear science*, pp. 3–35. New York: AIP.

Feller, W. (1956). *An introduction to probability theory and its applications*, Volume 1. New York: Wiley.

Field, R. J., E. Körös, and R. M. Noyes (1972). Oscillations in chemical systems. II. Thorough analysis of temporal oscillations in the bromate-cerium-malonic acid system. *J. Am. Chem. Soc. 94*, 8649–8664.

Field, R. J. and R. M. Noyes (1974). Oscillations in chemical systems. V. quantitative explanation of band migration in the Belousov-Zhabotinskii reaction. *J. Am. Chem. Soc. 96*, 2001–2006.

Fisher, R. A. (1937). The wave of advance of advantageous genes. *Ann. Eugenics 7*, 355–369.

FitzHugh, R. (1969). Mathematical models of excitation and propagation in nerve and muscle. In H. P. Schwan (Ed.), *Biological engineering*. New York: McGraw-Hill.

FitzHugh, R. A. (1961). Impulses and physiological states in theoretical models of nerve membrane. *Biophys. J. 1*, 445–466.

Flytzanis, N., E. Yiachnakis, and J. Micheloyannis (1991). Analysis of EEG signals and their spatial correlation over the scalp surface. In M. Markus and A. V. Holden (Eds.), *Nonlinear wave processes in excitable media*. New York: Plenum.

Fokas, A. S. and T. Bountis (1996). Order and the ubiquitous occurrence of chaos. *Physica A 228*(1-4), 236–244.

Fowler, A. C. and M. J. McGuinness (1982). Hysteresis in the Lorenz equations. *Phys. Lett. A 92*, 103.

Fowler, A. C., M. J. McGuinness, and J. D. Gibbon (1982). The complex Lorenz equations. *Physica D 4*, 139.

Frank, M., R. Gencay, and T. Stengos (1988). International chaos? *Europ. Econ. Rev. 32*, 1569–1584.

Frank, M. and T. Stengos (1988a). The stability of Canadian macroeconomic data as measured by the largest Lyapunov exponent. *Econ. Lett. 27*, 11–14.

Frank, M. Z. and T. Stengos (1988b). Chaotic dynamics in economic time-series. *J. Economic Surveys 2*, 103–133.

Frank, M. Z. and T. Stengos (1988c). Some evidence concerning macroeconomic chaos. *J. Monetary Economics 22*, 423–438.

Frederickson, P., J. L. Kaplan, E. D. Yorke, and Y. J. A. (1983). The Lyapunov dimension of strange attractors. *J. Diff. Eqns. 49*, 185–207.

Frouzakis, C. E., R. A. Adomaitis, and I. G. Kevrekidis (1991). Resonance phenomena in an adaptively-controlled system. *Int. J. of Bifurcation and Chaos 1*(1), 83–106.

Frouzakis, C. E., L. Gardini, I. G. Kevrekidis, G. Millerioux, and C. Mira (1997). On some properties of invariant sets of two-dimensional noninvertible maps. *Int. J. of Bifurcation and Chaos 7*(6), 1167–1194.

Fukuda, W. and S. Katsura (1986). Exactly solvable models showing chaotic behavior II. *Physica A 136*, 588.

Galias, Z. and P. Zgliczynski (1998). Computer assisted proof of chaos in the Lorenz equations. *Physica D 115*, 165–188.

Gallas, J. A. C. (1993). Structure of the parameter space of the Hénon map. *Phys. Rev. Lett. 70*(18), 2714–2717.

Garbaczewski, P., M. Wolf, and A. Weron (Eds.) (1995). *Chaos — The interplay between stochastic and deterministic behaviour. Proc. of the 31st Winter School of Theor Phys., Karpasz, Poland, 13.-24. Feb. 1995*. Berlin: Springer-Verlag.

Gause, G. F. and A. A. Witt (1934). On the periodic fluctuations in the numbers of animals. A mathematical theory of the relaxation interaction between predators and prey and its application to a population of protozoa. *Bull. Acad. Sci. U.R.S.S.*, 1551–1559.

Georgiou, I. T. (1999). On the global geometric structure of the dynamics of the elastic pendulum. *Nonlinear Dynamics 18*(1), 51–68.

Geysermans, P. and F. Baras (1997). Stochastic description of a period-2 limit cycle. *Europhys. Lett. 40*(1), 1–6.

Giovanis, A. and C. H. Skiadas (1995). Forecasting the electricity consumption by applying stochastic modeling techniques. In J. Janssen, C. H. Skiadas, and C. Zopounidis (Eds.), *Advances in stochastic modelling and data analysis*. Dordrecht: Cluwer Acad. Publishers.

Glass, L. (1975). Global analysis of nonlinear chemical kinetics. In B. Berne (Ed.), *Statistical mechanics*, pp. 311–349. New York: Plenum.

Glass, L. (1988). Simple mathematical models for complex dynamics in physiological systems. In B.-L. Hao (Ed.), *Directions in chaos*, Volume 2, pp. 90. Singapore: World Scientific.

Glass, L. (1991). Cardiac arrhythmias and circle maps. *Chaos 1*, 13–19.

Glass, L., M. R. Guevara, J. Belair, and A. Shrier (1984). Global bifurcations of a periodically forced biological oscillator. *Phys. Rev. A 29*, 1348–1357.

Glass, L. and P. Hunter (1990). There is A theory of heart. *Physica D 43*, 1–16.

Glass, L. and M. C. Mackey (1988). *From clocks to chaos*. Princeton NJ: Univ. Press.

Glass, L. and J. S. Pasternack (1978). Stable oscillations in mathematical models of biological control systems. *J. Math. Biol. 6*, 207–223.

Glass, L. and R. Young (1979). Structure and dynamics of neural network oscillators. *Brain Res. 179*, 207–218.

Glass, L. and W. Zeng (1990). Complex bifurcations and chaos in simple theoretical models of cardiac oscillations. In *Mathematical Approaches to Cardiac Arrhythmias*, Volume 591 of *Ann. N.Y. Acad. Sci.*, pp. 316–327.

Gleick, J. (1987). *Chaos — Making a new science*. New York: Viking.

Goetz, A. (1998). Perturbations of 8-attractors and births of satellite systems. *Int. J. of Bifurcation and Chaos 8*(10), 1937–1956.

Gollub, J. P. and H. L. Swinney (1975). Onset of turbulence in a rotating fluid. *Phys. Rev. Lett. 35*, 927.

Gompertz, B. (1825). On the nature of the function expressing of the law of human mortality. *Philosoph. Trans. Royal Soc. 36*, 513–585.

Goriely, A. (1992). From weak to full Painlevé property via time singularities transformations. In T. Bountis (Ed.), *Chaotic dynamics and practice*. London: Plenum.

Grammaticos, B., B. Dorizzi, and A. Ramani (1983). Integrability of Hamiltonians with third and fourth degree polynomial potentials. *J. Math. Phys. 24*, 2289.

Grammaticos, B., A. Ramani, and V. G. Papageorgiou (1991). Do integrable mappings have the Painlevé property? *Phys. Rev. Lett. 67*, 1825.

Grassberger, P. (1981). On the Hausdorff dimension of fractal attractors. *J. Stat. Phys. 26*, 173.

Grassberger, P. (1983a). Generalized dimensions of strange attractors. *Phys. Lett. A 97*, 227–230.

Grassberger, P. (1983b). On the fractal dimension of the Hénon attractor. *Phys. Lett. A 97*, 224–226.

Grassberger, P. (1985). Information flow and maximum entropy measures for 1-D maps. *Physica D 14*, 365–373.

Grassberger, P. (1986). Toward a quantitative theory of self-generated complexity. *Int. J. Theor. Phys. 25*(9), 907–938.

Grassberger, P. and I. Procaccia (1983a). Estimation of the Kolmogorov entropy from a chaotic signal. *Phys. Rev. A 28*, 2591–2593.

Grassberger, P. and I. Procaccia (1983b). Measuring the strangeness of strange attractors. *Physica D 9*, 189–208.

Grassberger, P. and I. Procaccia (1983c). On the characterization of strange attractors. *Phys. Rev. Lett. 50*, 346–349.

Grassberger, P. and M. Scheunert (1981). Some more universal scaling laws for critical mappings. *J. Stat. Phys. 26*, 697.

Grassberger, P. and T. Schreiber (1999). Statistical mechanics — microscopic chaos from Brownian motion? *Nature 401*(6756), 875–876.

Grau, M. and et al (1983). On the periodic orbits of the Contopoulos Hamiltonian. In L. Garrido (Ed.), *Dynamical systems and chaos*, Volume 179 of *Lecture notes in physics*, pp. 284–286. Berlin: Springer-Verlag.

Grebogi, C., E. Ott, S. Pelikan, and J. A. Yorke (1984). Strange attractors that are not chaotic. *Physica D 13*, 261–268.

Grebogi, C., E. Ott, and J. A. Yorke (1982). Chaotic attractors in crisis. *Phys. Rev. Lett. 48*, 1507–1510.

Grebogi, C., E. Ott, and J. A. Yorke (1987). Unstable periodic orbits and the dimension of chaotic attractors. *Phys. Rev. A 36*, 3522–3524.

Grebogi, C., E. Ott, and J. A. Yorke (1988). Roundoff induced periodicity and the correlation dimension of chaotic attractors. *Phys. Rev. A 38*, 366.

Guckenheimer, J. (1976). A strange strange attractor. In J. E. Marsden and M. McCracken (Eds.), *The Hopf bifurcation and applications*, pp. 368–381. Berlin: Springer-Verlag.

Guckenheimer, J. (1979a). The bifurcation of quadratic functions. *Annals of N.Y. Acad. Sc. 316*, 78–85.

Guckenheimer, J. (1979b). Sensitive dependence to initial conditions for one-dimensional maps. *Commun. Math. Phys. 70*, 133.

Guckenheimer, J. (1980a). Dynamics of the van der Pol equation. *IEEE Trans. Circ. and Systems CAS 27*, 983.

Guckenheimer, J. (1980b). Symbolic dynamics and relaxation oscillations. *Physica D 1*, 227–235.

Guckenheimer, J. and P. J. Holmes (1983). *Nonlinear oscillations, dynamical systems, and bifurcations of vector fields*. Berlin: Springer-Verlag.

Guckenheimer, J. and P. J. Holmes (1988). Structurally stable heteroclinic cycles. *Math. Proc. Cambridge Philos. Soc. 103*, 189–192.

Guevara, M. R., L. Glass, M. C. Mackey, and A. Shrier (1983). Chaos in neurobiology. *IEEE Trans. SMC 13*, 790.

Hadamard, J. (1898). Les surfaces à courbures opposées et leurs lignes géodésiques. *J. Math. Pure Appl. 4*, 27.

Haken, H. (1978). *Synergetics; An introduction*. New York: Springer-Verlag.

Haken, H. and G. Mayer-Kress (1981). Chapman-Kolmogorov equation and path integrals for discrete chaos in presence of noise. *Z. Physik 43*, 185–187.

Hao, B.-L. (Ed.) (1984). *Chaos*. Singapore: World Scientific.

Hao, B.-L. (Ed.) (1990). *Chaos II*. Singapore: World Scientific.

Hassard, B., S. P. Hastings, W. C. Troy, and J. Zhang (1994). A computer proof that the Lorenz equations have "chaotic" solutions. *Appl. Math. Lett., in press*.

Hausdorff, F. (1919). Dimension und äußeres maß. *Math. Annalen 79*, 157.

Helleman, R. H. G. (1980a). *Nonlinear dynamics*, Volume 357 of *Ann. N. Y. Acad. Sci.* New York: The New York Academy of Sciences.

Helleman, R. H. G. (1980b). Self-generated chaotic behavior in nonlinear mechanics. In E. G. D. Cohen (Ed.), *Fundamental problems in statistical mechanics*, Volume 5, pp. 165. Amsterdam: North-Holland.

Helleman, R. H. G. (1981). One mechanism for the onsets of largescale chaos in conservative and dissipative systems. In W. H. et al. (Ed.), *Nonequilibrium problems in statistical mechanics*, Volume 2. New York: Wiley.

Helleman, R. H. G. (1984). Feigenbaum sequences in conservative and dissipative systems. *Springer Ser. in Synerg. 11.*

Hénon, M. (1969). Numerical study of quadratic area-preserving mappings. *Quart. Appl. Math. 27*, 291–312.

Hénon, M. (1976). A two-dimensional mapping with a strange attractor. *Commun. Math. Phys. 50*, 69–77.

Hénon, M. (1982). On the numerical computation of Poincaré maps. *Physica D 5*, 412–414.

Hénon, M. (1989). Chaotic scattering modelled by an inclined billiard. *Physica D 33*, 132–156.

Hénon, M. and C. Heiles (1964). The applicability of the third integral of motion: Some numerical experiments. *Astron. J. 69*, 73–79.

Hénon, M. and Y. Pomeau (1977). Two strange attractors with a simple structure. In R. Temam (Ed.), *Turbulence and Navier-Stokes equations*, Volume 565 of *Lecture notes in mathematics*, pp. 29. Berlin: Springer-Verlag.

Hilborn, R. C. (1994). *Chaos and nonlinear dynamics: an introduction for scientists and engineers*. Oxford New York: Oxford Univ. Press.

Hirsch, M. W. and S. Smale (1974). *Differential equations, dynamical systems, and linear algebra*. New York: Academic Press.

Hodgkin, A. L. and A. F. Huxley (1952a). Current carried by sodium and potassium ions through the membrane of the giant axon of loligo. *J. Physiology 116*, 449–472.

Hodgkin, A. L. and A. F. Huxley (1952b). A quantitative description of membrane current and its application to conduction and excitation in nerve. *J. Physiol. 116*, 500–544.

Holmes, C. and P. J. Holmes (1981). Second order averaging and bifurcations to subharmonics in Duffing's equation. *J. Sound Vib. 78*, 161–174.

Holmes, P. J. (1977a). Behavior of an oscillator with even nonlinear damping. *Int. J. Nonlinear Mech. 12*, 323–326.

Holmes, P. J. (1977b). Strange phenomena in dynamical systems and their physical implications. *Applied Mathematical Modelling 7*(1), 362–366.

Holmes, P. J. (1979a). A nonlinear oscillator with a strange attractor. *Phil. Trans. Roy. Soc. London 292*, 419–448.

Holmes, P. J. (1979b). Recurrent periodic and nonperiodic behavior in simple dynamical systems. In *Proc. Joint National Meeting*, pp. 107.

Holmes, P. J. (1980). Periodic, non-periodic and irregular motions in a Hamiltonian system. *Rocky Mountain J. Math 10*(4), 679–693.

Holmes, P. J. (1982). Proof of non-integrability for the Hénon-Heiles Hamiltonian near an exceptional integrable case. *Physica D 5*, 35–347.

Holmes, P. J. (1984). Bifurcation sequences in horseshoe maps: infinitely many routes to chaos. *Phys. Lett. 104A*, 299–302.

Holmes, P. J. (1986). Spatial structure of time-periodic solutions of the Ginzburg-Landau equation. *Physica D 23*, 4–90.

Holmes, P. J. (1990). Poincaré, celestial mechanics, dynamical systems theory and "chaos". *Physics Reports 193*(3), 137–163.

Holmes, P. J. and J. E. Marsden (1982). Horseshoes in perturbations of Hamiltonian systems with two degrees of freedom. *Comm. Math. Phys. 82*, 523–544.

Holmes, P. J. and D. A. Rand (1976). The bifurcations of Duffing's equation: An application of catastrophy theory. *J. Sound Vib. 44*(2), 237–253.

Holmes, P. J. and D. A. Rand (1978). Bifurcation of the forced van der Pol oscillator. *Quart. Appl. Math. 35*, 495–509.

Hopf, E. (1942). Abzweigung einer periodischen Lösung von einer stationären Lösung eines differentialsystems. *Ber. Math.-Phys. Akad. Wiss. Leipz. 94*, 3–22.

Hopf, E. (1948). A mathematical example displaying features of turbulence (excerpt). *Commun. Pure Appl. Math. 1*, 303.

Hopf, F. A., D. L. Kaplan, H. M. Gibbs, and R. L. Shoemaker (1982). Bifurcations to chaos in optical bistability. *Phys. Rev. A 25*, 2172.

Hoppensteadt, F. C. (1993). *Analysis and simulation of chaotic systems*. New York: Springer-Verlag.

Hudson, J. L., O. E. Rössler, and H. C. Killory (1988). Chaos in a four-variable piecewise-linear system of differential equations. *IEEE Trans. Circ. and Systems CAS 35*, 902–908.

Hunter, C. and A. Toomre (1969). Dynamics of the bending of the galaxy. *Astrophys. J. 155*, 747.

Ide, K. and S. Wiggins (1989). The bifurcation to homoclinic tori in the quasiperiodically forced Duffing oscillator. *Physica D 34*, 169–182.

Ikeda, K. (1979). Multiple-valued stationary state and its instability of the transmitted light by a ring cavity system. *Opt. Commun. 30*, 257.

Ikeda, K. and O. Akimoto (1982). Instability leading to periodic and chaotic self-pulsations in a bistable optical cavity. *Phys. Rev. Lett. 48*, 617.

Ikeda, K., H. Daido, and O. Akimoto (1980). Optical turbulence: Chaotic behavior of transmitted light from a ring cavity. *Phys. Rev. Lett. 45*, 709.

Ikeda, K. and K. Kondo (1982). Successive higher-harmonic bifurcations in systems with delayed feedback. *Phys. Rev. Lett. 49*, 1467.

Ikeda, K. and K. Matsumoto (1987). High-dimensional chaotic behavior in systems with time-delayed feedback. *Physica D 29*, 223–235.

Ikeda, T. and S. Murakami (1999). Dynamic response and stability of a rotating asymmetric shaft mounted on a flexible base. *Nonlinear Dynamics 20*(1), 1–19.

Ikeda, Y. and S. Tokinaga (2000). Approximation of chaotic dynamics by using smaller number of data based upon the genetic programming and its applications. *IEICE Trans. Fund. Elec. Com. Com. E83A*(8), 1599–1607.

Iooss, G., R. Helleman, and R. Stora (Eds.) (1983). *Chaotic behavior of deterministic systems*. Amsterdam: North Holland.

Ishida, Y., T. Ikeda, T. Yamamoto, and S. Murakami (1989). Nonstationary vibration of a rotating shaft with nonlinear spring characteristics during acceleration through a critical speed (A critical speed of a 1/2-order subharmonic oscillation). *JSME Int. J. Ser. III 32*(4), 575–584.

Ito, A. (1979). Successive subharmonic bifurcations and chaos in a nonlinear Mathieu equation. *Prog. Theor. Phys. 61*, 815.

Ito, H. and L. Glass (1991). Spiral breakup in a new model of discrete excitable media. *Phys. Rev. Lett. 66*, 671–674.

Itoh, H. (1985). Nonintegrability of Hénon-Heiles system and a theorem of ziglin. *Kodai Math. J. 8*, 120.

Joy, M. P. (1995). On the bifurcation in a modulated logistic map: Comment. *Phys. Lett. A 202*(2-3), 237–239.

Julia, G. (1918). Mémoire sur l'itération des fonctions rationelles. *J. Math. Pure Appl. 8*, 47–245.

Kac, M. (1947). Random walk and the theory of Brownian motion. *Am. Math. Month. 54*(7), 369–391.

Kadanoff, L. P. (1981). Scaling for a critical Kolmogorov-Arnold-Moser trajectory. *Phys. Rev. Lett. 47*, 1641–1643.

Kadanoff, L. P. (1993). *From order to chaos: Essays: Critical, chaotic and otherwise*. Singapore: World Scientific.

Kahlert, C. and O. E. Rössler (1984). Chaos as a limit in a boundary value problem. *Z. Naturforsch. A 39*, 1200–1203.

Kahlert, C. and O. E. Rössler (1985). Analytical properties of Poincaré half-maps in a class of piecewise-linear dynamical systems. *Z. Naturforsch. 40*, 1011–1025.

Kahlert, C. and O. E. Rössler (1986). The separating mechanisms in Poincaré half-maps. *Z. Naturforsch. A 41*, 1369–1380.

Kahlert, C. and O. E. Rössler (1987). Analogues to a boundary away from analyticity. *Z. Naturforsch. 42*, 324–328.

Kalman, R. E. (1956). Nonlinear aspects of sample data control systems. In *Proc. Symp. on Nonlin. Circ. Anal., Brooklyn, April 1956*, pp. 273–313.

Kalman, R. E. (1960). A new approach to linear filtering and prediction problems. *J. Basic Eng. 82*, 35–45.

Kapitaniak, T. (1990). Analytical condition for chaotic behavior of the Duffing oscillator. *Phys. Lett. A 144*, 322–324.

Kapitaniak, T. (1991). On strange non-chaotic attractors and their dimension. *Chaos Solitons and Fractals 1*, 67–77.

Kaplan, H. (1983). New method for calculating stable and unstable periodic orbits of one-dimensional maps. *Phys. Lett. A 97*, 365.

Kaplan, J. L. and J. A. Yorke (1979a). Chaotic behavior of multidimensional difference equations. In H. O. Walter and H.-O. Peitgen (Eds.), *Functional differential equations and approximation of fixed points*, Volume 730 of *Lecture notes in mathematics*, pp. 204–227. Berlin: Springer-Verlag.

Kaplan, J. L. and J. A. Yorke (1979b). *The onset of chaos in a fluid flow model of Lorenz*, Volume 316 of *Ann. N. Y. Acad. Sci.* New York: The New York Academy of Sciences.

Kappos, E. (1996). The Conley index and global bifurcations. II. Illustrative applications. *J. of Bifurcation and Chaos 6*(12B), 2491–2505.

Karamanos, K. and G. Nicolis (1999). Symbolic dynamics and entropy analysis of Feigenbaum limit sets. *Chaos Solitons Fractals 10*(7), 1135–1150.

Károlyi, G., I. Scheuring, and T. Czárán (2002). Metabolic network dynamics in open chaotic flow. *Chaos 12*(2), 460–469.

Károlyi, G. and T. Tél (1997). Chaotic tracer scattering and fractal basin boundaries in a blinking vortex-sink system. *Physics Reports 290*, 125–147.

Katsamaki, A. and C. H. Skiadas (1995). Analytic solution and estimation of parameters on a stochastic exponential model for a technological diffusion process. *Appl. Stoch. Mod. Dat. Analys. 11*, 59–75.

Katz, L. (1996). Note on the Schwarzian derivative. *Chaos Solitons Fractals 7*(9), 1495–1496.

Keppenne, C. and C. Nicolis (1988). Toward a quantitative view of predictability of weather. *Annal. Geophys.*, 207.

Keppenne, C. L. and C. Nicolis (1989). Global properties and local structure of the weather attractor over western Europe. *J. Atmos. Sci. 46*, 2356.

Kerr, R. A. (1989). Does chaos permeate the solar system? *Science 244*, 144–45.

Kevrekidis, I. G., R. Aris, L. D. Schmidt, and S. Pelikan (1985). Numerical computation of invariant circles of maps. *Physica D 16*, 243–251.

Kevrekidis, I. G. and M. S. Jolly (1987). On the use of interactive graphics in the numerical study of chemical dynamics. *1987 AIChE Annual Meeting, New York 22*.

Kevrekidis, I. G., R. Rico-Martínez, R. E. Ecke, R. M. Farber, and A. S. Lapedes (1994). Global bifurcations in Raleigh-Bénard convection. Experiments, empirical maps and numerical bifurcation analysis. *Physica D 71*, 342–362.

Kilias, T. (1994). Generation of pseudo-chaotic sequences. *Int. J. of Bifurcation and Chaos 4*(3), 709–713.

Kim, J. H. and J. Stringer (Eds.) (1997). *Applied chaos.* New York: Wiley.

Kirner, T. and O. E. Rössler (1997). Frequency clustering in a chain of weakly coupled oscillators. *Z. Naturforsch. A 52*(8-9), 578–580.

Kloeden, P. E. and J. Lorenz (1986). Stable attracting sets in dynamical systems and in their one-step discretizations. *SIAM J. Numer. Anal. 23*, 986–995.

Kloeden, P. E. and E. Platen (1989). A survey of numerical methods for stochastic differential equations. *J. Stoch. Hydrol. Hydraul. 3*, 155–178.

Kloeden, P. E. and E. Platen (1992). *The numerical solution of stochastic differential equations.* Berlin: Springer-Verlag.

Kloeden, P. E., E. Platen, and H. Schurz (1991). The numerical solution of nonlinear stochastic dynamical systems: A brief introduction. *Int. J. of Bifurcation and Chaos 1*(2), 277–286.

Kocak, H. (1989). *Differential and difference equations through computer experiments.* Berlin: Springer-Verlag.

Kolmogorov, A. (1942). Interpolation and extrapolation of stationary series. *Bulletin de l'Académie des Sciences de l'U.R.S.S. Series Mathématiques 2*, 3.

Kolmogorov, A., I. Petrovsky, and N. Piscounoff (1937). *Etude de l'équation de la diffusion avec croissance de la quantitéde matière et son application àun problème biologique*, Volume 1. Bull. Univ. Moskou: Ser. internat.

Kolmogorov, A. N. (1936). Sulla teoria di Volterra della lotta per l'esistenza, (in Italian). *Giorn. Istituto Ital. d. Attuari 7*, 74–80.

Kolmogorov, A. N. (1954). Preservation of conditionally periodic movements with small change in the Hamiltonian function. *Dokl. Akad. Nauk. SSSR 98*, 527.

Kolmogorov, A. N. (1958). A new invariant of transitive dynamical systems. *Dokl. Akad. Nauk. SSSR 119*, 861.

Kolmogorov, A. N. (1965). Three approaches to the quantitative definition of information. *Inf. Trans. 1*, 3–11.

Komineas, S., M. N. Vrahatis, and T. Bountis (1994). 2D universality of period-doubling bifurcations in 3D conservative reversible mappings. *Physica A 211*(2-3), 218–233.

Kopell, N. and L. N. Howard (1973a). Horizontal bands in the Belousov reaction. *Science 180*, 1171–1173.

Kopell, N. and L. N. Howard (1973b). Plane wave solutions to reaction-diffusion equations. *Studies in Appl. Math. 52*, 291–328.

Kosmatopoulos, E. B., M. M. Polycarpou, M. A. Christodoulou, and P. A. Ioannou (1995). High order neural network structures for identification of dynamical systems. *IEEE Trans. Neural Networks 6*, 422–431.

Koumoutsos, N. and C. H. Skiadas (1995). Applied stochastic models and data analysis for engineering education. In J. Janssen, C. H. Skiadas, and C. Zopounidis (Eds.), *Advances in stochastic modelling and data analysis*. Dordrecht: Cluwer Acad. Publishers.

Kounadis, A. N. (1999). Dynamic buckling of autonomous systems having potential energy universal unfoldings of cuspoid catastrophe. *Nonlinear Dynamics 18*(3), 235–252.

Kowalevskaya, S. (1889a). Sur le problème de la rotation d'un corps solide autour d'un point fixe. *Acta Math. 12*, 177.

Kowalevskaya, S. (1889b). Sur une propriété du système d'équations différentielles qui définit la rotation d'un corps solide autour d'un point fixe. *Acta Math. 13*, 81.

Kremmydas, G. P., A. V. Holden, A. Bezerianos, and T. Bountis (1996). Representation of sino-atrial node dynamics by circle maps. *Int. J. of Bifurcation and Chaos 6*(10), 1799–1805.

Kuczma, M. (1968). *Functional equations in a single variable*. Warszawa: PWN - Polish Scientific Publishers, Warszawa.

Kugiumtzis, D. (1997). Correction of the correlation dimension for noisy time series. *Int. J. of Bifurcation and Chaos 7*(6), 1283–1294.

Kuske, R. and G. Papanicolaou (1998). The invariant density of a chaotic dynamical system with small noise. *Physica D 120*(3-4), 255–272.

Kutta, W. (1901). Beitrag zur näherungsweisen integration totaler differentialgleichungen. *Z. Math. Phys. 46*, 435–453.

Kuznetsov, S. P., A. S. Pikovsky, and U. Feudel (1995). Birth of strange nonchaotic attractor: a renormalization group analysis. *Phys. Rev. E 51*(3, pt. A), 1629–1632.

Kuznetsov, Y. A. (2004). *Elements of applied bifurcation theory*. New York: Springer-Verlag.

Kuznetsov, Y. A., S. Muratori, and S. Rinaldi (1992). Bifurcations and chaos in a periodic predator-prey model. *Int. J. of Bifurcation and Chaos 2*(1), 117–128.

Lamb, H. (1879). *A Treatise on the Mathematical Theory of the Motion of Fluids*. University Press.

Landau, L. D. (1944). On the problem of turbulence, (Russ. original); English translation in. *Akad. Nauk. Doklady 44*, 339.

Landberg, P. (1956). Vibrations caused by chip formation. *Microtenic 10*, 219–228.

Lanford, O. E. (1982a). A computer assisted proof of the Feigenbaum conjectures. *Bull. Am. Math. Soc. 6*, 427–434.

Lanford, O. E. (1982b). The strange attractor theory of turbulence. *Ann. Rev. of Fluid Mech. 14*, 347.

Lanford, O. E. (1983). Period doubling in one and several dimensions. *Physica D 7*, 124–125.

Lanford, O. E. (1984). Functional equations for circle homeomorphisms with golden ratio rotation number. *J. Stat. Phys. 34*, 57–73.

Lasota, A. and M. C. Mackey (1985). *Probabilistic properties of deterministic systems*. Cambridge: University Press.

Lasota, A. and M. C. Mackey (1994). *Chaos, Fractals, and Noise (Applied Mathematical Sciences)*, Volume 97. Berlin: Springer-Verlag.

Lauterborn, W. (1976). Numerical investigation of nonlinear oscillations of gas bubbles in liquids. *J. Acoust. Soc. Am. 59*, 283–293.

Leith, C. E. (1996). Stochastic models of chaotic systems. *Physica D 98*(2-4), 481–491.

Leonov, N. N. (1959). Transformations d'une droite en elle-même. *Radiofiscia 2*, 942.

Leonov, N. N. (1960a). Théorie des transformations discontinues d'une droite en elle-même. *Radiofiscia 3*, 872.

Leonov, N. N. (1960b). Transformation ponctuelle d'une droite en elle-même discontinue, linéaire par morceaux. *Radiofiscia 3*, 496.

Leontaritis, I. J. and S. A. Billings (1987). Experimental design and identificability for nonlinear systems. *Int. J. Syst. Sci. 18*(1), 189–202.

Lewis, J. E. and L. Glass (1991). Steady states, limit cycles, and chaos in models of complex biological networks. *Int. J. of Bifurcation and Chaos 1*(2), 477–483.

Li, K. T. and L. Z. Wang (2001). Global bifurcation and long time behavior of the Volterra-Lotka ecological model. *Int. J. Bifurcation Chaos 11*(1), 133–142.

Li, T. Y. and J. A. Yorke (1975). Period three implies chaos. *Am. Math. Monthly 82*, 985–992.

Lorenz, E. (1993). *The essence of chaos.* Seattle: Univ. of Washington Press.

Lorenz, E. N. (1956). Empirical orthonogal functions and statistical weather prediction. Scientific Report No. 1, Statistical Forecasting Project, Cambridge, MA.

Lorenz, E. N. (1963). Deterministic nonperiodic flow. *J. Atmos. Sci. 20*, 130–141.

Lorenz, E. N. (1964). The problem of deducing the climate from the governing equations. *Tellus 16*, 1.

Lorenz, E. N. (1969). Atmospheric predictability as revealed by naturally occurring analogies. *J. Atmos. Sci. 26*, 636.

Lorenz, E. N. (1979). Predictability: Does the flap of a butterfly's wings in Brazil set off a tornado in Texas? Talk given at the annual meeting of the AAAS December 29, 1979 in Washington, American Association for the Advancement of Science.

Lorenz, E. N. (1980). Noisy periodicity and reverse bifurcation. *Ann. N. Y. Acad. Sci. 357*, 282–291.

Lorenz, E. N. (1991). Dimension of weather and climate attractors. *Nature 353*, 241–244.

Lotka, A. (1910). Zur theorie der periodischen raktionen. *Z. phys. Chemie 72*, 508.

Lozi, R. and S. Ushiki (1988). Organized confinors and anticonfinors and their bifurcations in constrained. *Ann. Tèlècommun. 43*, 187–208.

Lozi, R. and S. Ushiki (1991a). Coexisting chaotic attractors in Chua's circuit. *Int. J. of Bifurcation and Chaos 1*(4), 923–926.

Lozi, R. and S. Ushiki (1991b). Confinors and bounded-time patterns in Chua's circuit and the double scroll family. *Int. J. of Bifurcation and Chaos 1*(1), 119–138.

Lyapunov, M. A. (1893). in Russian, for an English translation see: Ann. Math. Studies 17, Princeton (1949); cited after W. Hahn (1959) "Theorie und anwendung der direkten methode von Lyapunov". *Mat. Sbornik 17*, 252–333.

Lyapunov, M. A. (1947). Problème général de la stabilité du mouvement, (1907 translation from the 1892 Russian original). *Annal. Math. Stud. 17*.

Lynden-Bell, D. (1969). Galactic nuclei as collapsed old quasars. *Nature 223*, 690.

MacKay, R. S. (1993). Non-area-preserving directions from area-preserving golden circle fixed points. *Nonlinearity. 6*(5), 799.

MacKay, R. S. (2001). Complicated dynamics from simple topological hypotheses. *Philos. Trans. R. Soc. Lond. Ser. A Math. Phys. Eng. Sci. 359*(1784), 1479–1496.

Mackey, M. C. and L. Glass (1977). Oscillation and chaos in physiological control systems. *Science 197*, 287–289.

Mahajan, V. and M. Schoeman (1977). Generalized model for the time pattern of the diffusion process. *IEEE Transactions on Engineering Management 24*, 12–18.

Malliaris, A. G. and J. L. Stein (1999). Methodological issues in asset pricing: Random walk or chaotic dynamics. *J. Banking and Finance 23*(11), 1605–1635.

Malthus, T. (1798). *An essay on the principle of population.* London: Johnson.

Mandelbrot, B. (1963). The variation of certain speculative prices. *J. Business 36*, 394–413.

Mandelbrot, B. (1967). The variation of some other speculative prices. *J. Business 40*, 393–413.

Mandelbrot, B. (1972). Statistical methodology for nonperiodic cycles: From the covariance to R/S analysis. *Ann. Econ. Soc. Measurement 1*.

Mandelbrot, B. (1977). *Fractals — form, chance, and dimension.* San Francisco: Freeman.

Mandelbrot, B. (1980). Fractal aspects of the iteration of $z \mapsto \lambda * z(1 - z)$ for complex λ and z. *Ann. N. Y. Acad. Sci. 357*, 249–259.

Mandelbrot, B. (1982). *The fractal geometry of nature.* San Francisco: Freeman.

Mandelbrot, B. and J. Wallis (1969a). Computer experiments with fractional Gaussian noises. Part 1. Averages and variances. *Water Resourc. Res. 5*.

Mandelbrot, B. and J. Wallis (1969b). Computer experiments with fractional Gaussian noises. Part 2. Rescaled ranges and spectra. *Water Resourc. Res. 5*.

Mandelbrot, B. and J. Wallis (1969c). Computer experiments with fractional Gaussian noises. Part 3. Mathematical appendix. *Water Resourc. Res. 5*.

Mandelbrot, B. B. (1974). Intermittent turbulence in self-similar cascades: Divergence of high moments and dimension of the carrier. *J. Fluid Mech. 62*, 331.

Mandelbrot, B. B. (1990). Two meanings of multifractality, and the notion of negative fractal dimension. In D. K. Campbell (Ed.), *Chaos — Soviet-American perspectives on nonlinear science*, pp. 79–90. New York: AIP.

Mandelbrot, B. B. and J. V. Ness (1968). Fractional Brownian motions, fractional noises and applications. *SIAM Review 10*, 422–437.

Manneville, P. (1980). Intermittency, self-similarity and $1/f$-spectrum in dissipative dynamical systems. *J. d. Phys. 41*, 1235–1241.

Manneville, P. (1989). *Dissipative structures and weak turbulence*. New York: Academic Press.

Manneville, P. and Y. Pomeau (1979). Intermittency and the Lorenz model. *Phys. Lett. A 75*, 1.

Manneville, P. and Y. Pomeau (1980). Different ways to turbulence in dissipative dynamical systems. *Physica D 1*, 219–226.

Markstein, G. H. (1951). Experimental and theoretical studies of flame-front stability. *J. Aeronaut. Sci. 18*, 199–209.

Markus, M. and S. K. (1994). Observation of chemical turbulence in the Belousov - Zhabotinsky reaction. *Int. J. of Bifurcation and Chaos 4*(5), 1233–1243.

Marotto, F. R. (1979). Chaotic behavior in the Hénon mapping. *Commun. Math. Phys. 68*, 187.

Marsden, J. and M. McCracken (1975). *The Hopf bifurcation*. Berlin: Springer-Verlag.

Matsumoto, T. (1984). A chaotic attractor from Chua's circuit. *IEEE Trans. Circ. and Systems CAS 31*, 1055–1058.

May, R. M. (1972). Limit cycles in predator-prey communities. *Science 17*, 900–902.

May, R. M. (1974). Biological populations with nonoverlapping generations: Stable points, stable cycles, and chaos. *Science 186*, 645–647.

May, R. M. (1976). Simple mathematical models with very complicated dynamics. *Nature 261*, 459–467.

May, R. M. (1987a). Chaos and the dynamics of biological populations. In M. V. Berry, I. C. Percival, and N. O. Weiss (Eds.), *Proc. Royal Soc. Lond. A*, Volume 413, Princeton, NJ, pp. 27–44. University Press.

May, R. M. (1987b). When two and two make not four: nonlinear phenomena in ecology. *Proc. Roy. Soc. London A 413*, 27–44.

May, R. M. and G. F. Oster (1980). Period doubling and the onset of turbulence. An analytic estimate of the Feigenbaum ratio. *Phys. Lett. A 78*, 1.

Mayer-Kress, G. and H. Haken (1981). The influence of noise on the logistic model. *Journ. Stat. Phys. 26*, 149–171.

McKean, H. P. (1970). Nagumo's equation. *Adv. in Math. 4*, 209–223.

Meakin, P. (1986). A new model for biological pattern formation. *J. Theor. Biol. 118*, 101.

Meiss, J. D. (1999). Physics of chaos in Hamiltonian systems. *Nature 398*(6725), 303.

Metropolis, N., A. Rosenblut, M. Rosenbluth, A. Teller, and E. Teller (1953). Equations of state calculations by fast computing machine. *J. Chem. Phys. 21*, 1097.

Metropolis, N., M. L. Stein, and P. R. Stein (1973). On finite limit sets for transformations on the unit interval. *J. of Combinatorial Theory 15*(1), 25–44.

Michelson, D. (1986). Steady solutions of the Kuramoto-Sivashinsky equation. *Physica D 19*, 89–111.

Michielin, O. and P. E. Phillipson (1997). Map dynamics study of the Lorenz equations. *J. of Bifurcation and Chaos 7*(2), 373–382.

Misiurewicz, M. and B. Szewc (1980). Existence of a homoclinic point for Hénon map. *Commun. Math. Phys. 75*, 285.

Mitsi, S., S. Natsiavas, and I. Tsiafis (1998). Dynamics of nonlinear oscillators under simultaneous internal and external resonances. *Nonlinear Dynamics 16*(1), 23–39.

Modis, T. (1997). Genetic re-engineering of corporations. *Technol. Forecast. Social Change 56*, 107–118.

Modis, T. and A. Debecker (1992). Chaos-like states can be expected before and after logistic growth. *Technol. Forecast. Social Change 41*, 111–120.

Murakami, C., W. Murakami, and K. Hirose (2002, July). Sequence of global period doubling bifurcation in the Hénon map. *Chaos, Solitons and Fractals 14*(1), 1–17.

Muzzio, F. J. and J. M. Ottino (1990). Diffusion and reaction in a lamellar system: Self-similarity with finite rates of reaction. *Phys. Rev. A 42*, 5873–5884.

Myrberg, P. J. (1958). Iteration der reellen Polynome zweiten grades. *Ann. Acad. Sci. Fenn. 256*, 1–10.

Myrberg, P. J. (1962). Sur l'itération des polynomes réels quadratiques. *J. de Math. Pures et Appl. 9*, 339.

Nagumo, J., S. Arimoto, and S. Yoshizawa (1962). An active pulse transmission line simulating nerve axon. In *Proc. IRE*, Volume 50, pp. 2061–2070.

Namba, T. (1986). Bifurcation phenomena appearing in the Lotka-Volterra competition equation: A numerical study. *Math. Biosci. 81*, 191–212.

Naschie, M. S. E. (1989). Generalized bifurcation and shell buckling as special statical chaos. *ZAMM 69*, 376–377.

Naschie, M. S. E. and S. A. Athel (1989). On the connection between statical and dynamical chaos. *Z. Naturforsch. 44*, 645–650.

Naschie, M. S. E. and T. Kapitaniak (1990). Soliton chaos models for mechanical and biological elastic chains. *Phys. Lett. A 147*, 275–281.

Natsiavas, S., S. Theodossiades, and I. Goudas (2000). Dynamic analysis of piece-wise linear oscillators with time periodic coefficients. *Int. J. Non Lin. Mech. 35*(1), 53–68.

Natsiavas, S. and G. Verros (1999). Dynamics of oscillators with strongly nonlinear asymmetric damping. *Nonlinear Dynamics 20*(3), 221–246.

Navarro, J. F., C. S. Frenk, and D. M. White (1995). Simulations of X-ray clusters. *Mon. Not. R. Astron. Soc. 275*, 720.

Neimark, Y. I. and P. S. Landa (1992). *Stochastic and chaotic oscillations*. Dordrecht, Netherlands: Kluwer Academic.

Nicolis, C. and G. Nicolis (1984). Is there a climatic attractor? *Nature 311*, 529–532.

Nicolis, C. and G. Nicolis (1987). Evidence for climatic attractors. *Nature 326*, 523.

Nicolis, C. and G. Nicolis (1991). Dynamics for error growth in unstable systems. *Phys. Rev. A 43*, 5720–5723.

Nicolis, C. and G. Nicolis (1993). Finite time behavior of small errors in deterministic chaos and Lyapunov exponents. *Int. J. of Bifurcation and Chaos 3*(5), 1339–1342.

Nicolis, C. and G. Nicolis (1998). Closing the hierarchy of moment equations in nonlinear dynamical systems. *Phys. Rev. E 58*(4), 4391–4400.

Nicolis, G. (1971). Stability and dissipative structures in open systems far from equilibrium. *Adv. Chem. Phys. 19*, 209–324.

Nicolis, G. (1996). Natural laws and the physics of complex systems. In P. Wein-gartner and G. Schurz (Eds.), *Law and prediction in the light of chaos research*, Volume LNP 473, pp. 36–39. Berlin: Springer-Verlag.

Nicolis, G., T. Erneux, and M. Herschkowitz-Kaufman (1978). Pattern formation in reacting and diffusing systems. *Adv. Chem. Phys. 38*, 263–314.

Nicolis, G. and J. Portnow (1973). Chemical oscillations. *Chem. Rev. 73*, 365–384.

Nicolis, G. and I. Prigogine (1977). *Self-organization nonequilibrium systems*. New York: Wiley-Interscience.

Nicolis, G. and I. Prigogine (1981). Symmetry-breaking and pattern selection in far-from-equilibrium systems. In *Proc. Natl. Acad. Sci. USA*, Volume 78, pp. 659–663.

Nicolis, J. S., G. Mayer-Kress, and G. Haubs (1983). Non-uniform chaotic dynamics with implications to information processing. *Z. Naturforsch. 38a*, 1157–1169.

Nikolaev, E. V., V. N. Biktashev, and A. V. Holden (1999). On bifurcations of spiral waves in the plane. *Int. J. of Bifurcation and Chaos 9*(8), 1501–1516.

Nusse, H. E. and J. A. Yorke (1996). Basins of attraction. *Science 271*(5254), 1376–

1380.

Oono, Y. (1978). Period $\neq 2^n$ implies chaos. *Prog. Theor. Phys. 59*, 1028–1030.

Ornstein, D. S. (1995). In what sense can a deterministic system be random? *Chaos Solitons Fractals 5*(2), 139–141.

Ostriker, J. P. and E. P. J. Peebles (1973). A numerical study of the stability of flattened galaxies: Or, can cold galaxies survive? *Astrophys. J. 186*, 467–480.

Ott, E. (1981). Strange attractors and chaotic motions of dynamical systems. *Rev. Mod. Phys. 53*(4), 655–671.

Ott, E. (1993). *Chaos in dynamical systems.* Cambridge: University Press.

Ott, E., C. Grebogi, and J. A. Yorke (1989). Theory of first order phase transitions for chaotic attractors of nonlinear dynamical systems. *Phys. Lett. A 135*, 343–348.

Ott, E., C. Grebogi, and J. A. Yorke (1990). Controlling chaos. *Phys. Rev. Lett. 64*, 1196–1199.

Ott, E., T. Sauer, and J. A. Yorke (1994a). *Coping with chaos.* New York: Wiley.

Ott, E., T. Sauer, and J. A. Yorke (Eds.) (1994b). *Coping with chaos; Analysis of chaotic data and the exploitation of chaotic systems.* New York: Wiley & Sons Inc.

Ott, E., E. D. Yorke, and J. A. Yorke (1985). A scaling law. How an attractor's volume depends on noise level. *Physica D 16*, 62–78.

Ottino, J. M. (1989). *The kinematics of mixing: Stretching, chaos, and transport.* Cambridge: University Press.

Ottino, J. M. (1990). Mixing, chaotic advection, and turbulence. *Ann. Rev. Fluid Mech. 22*, 207–253.

Ozaki, T. (1982). The statistical analysis of perturbed limit cycle processes using nonlinear time series models. *J. Time Series Anal. 3*, 29.

Paidoussis, M. P., G. X. Li, and F. C. Moon (1989). Chaotic oscillations of the autonomous system of a constrained pipe conveying fluid. *J. Sound Vib. 135*, 1–19.

Palis, J. (2002). Chaotic and complex systems. *Curr. Sci. 82*(4), 403–406.

Panas, E. and V. Ninni (2000). Are oil markets chaotic? A non-linear dynamic analysis. *Energy Econ. 22*(5), 549–568.

Papadimitriou, S., T. Bountis, S. Mavroudi, and A. Bezerianos (2001). A probabilistic symmetric encryption scheme for very fast secure communication based on chaotic systems of difference equations. *Int. J. Bifurcation Chaos 11*(12), 3107–3115.

Papaioannou, G. and A. Karytinos (1995). Nonlinear time series analysis of the stock exchange: The case of an emerging market. *Int. J. of Bifurcation and Chaos 5*(6), 1557–1584.

Papoulias, F. A. (1991). Bifurcation analysis of line of sight vehicle guidance using sliding modes. *Int. J. of Bifurcation and Chaos 1*(4), 849–865.

Papoulias, F. A. and M. M. Bernitsas (1988). Autonomous oscillations, bifurcations and chaotic response of moored vessels. *J. Ship Res. 32*(3), 220–228.

Pasta, J. R., S. M. Ulam, and E. Fermi (1965). Studies on nonlinear problems. In *Collected works of Enrico Fermi*, Volume 2, pp. 978. Chicago: Univ. of Chicago Press.

Patsis, P. A., C. Efthymiopoulos, G. Contopoulos, and N. Voglis (1997). Dynamical spectra of barred galaxies. *Astron. Astrophys. 326*, 493–500.

Pearl, R. and L. J. Reed (1920). On the rate of growth of the population of the United States since 1790 and its mathematical representation. *Proceedings of the National Academy of Sciences of the United States of America 6*(6), 275–288.

Pecora, L. M. (Ed.) (1993). *Chaos in communications*, Volume 2038 of *SPIE proceedings*. Bellingham, Washington: SPIE-The International Society for Optical Engineering.

Pecora, L. M. and T. L. Carroll (1990). Synchronization in chaotic systems. *Phys. Rev. Lett. 64*, 821–824.

Pecora, L. M. and T. L. Carroll (1991). Driving systems with chaotic signals. *Phys. Rev. A 44*, 2374–2383.

Pecora, L. M. and T. L. Carroll (1993). Using chaos to eliminate multiple period behavior. *Phys. Rev. A 48*, 2426.

Pedersen, P. O. (1935). Subharmonics in forced oscillations in dissipative systems. part I. *J. Acoust. Soc. Am. 6*, 227–238.

Perez, R. and L. Glass (1982). Bistability, period doubling bifurcations and chaos in a periodically forced oscillator. *Phys. Lett. A 90*, 441.

Petit, J. M. and M. Hénon (1986). Satellite encounter. *Icarus 60*, 536.

Phatak, S. C. and S. S. Rao (1995). Logistic map: A possible random-number generator. *Phys. Rev. E 51*(4), 3670–3678.

Poincaré, H. (1890a). Mémoire sur les courbes définies par les équations différentielles I-VI, oeuvre I. In *Oeuvre I*. Paris: Gauthier-Villars.

Poincaré, H. (1890b). Sur les équations de la dynamique et le problème de trois corps. *Acta Math. 13*, 1–270.

Poincaré, H. (1892). *Les méthodes nouvelles de la mécanique céleste*. Paris:

Gauthier-Villars.

Poincaré, H. (1899). *Les méthodes nouvelles de la mécanique céleste*, Volume 3. Paris: Gauthier-Villars.

Poincaré, H. (1905). *Leçons de mécanique*. Paris: Gauthier-Villars.

Poincaré, H. (1908). *Science et methode*. Biblioteque Scientifique.

Pomeau, Y. and P. Manneville (1980). Intermittent transition to turbulence in dissipative dynamical systems. *Commun. Math. Phys. 74*, 189–197.

Poznanski, K. Z. (1983). International diffusion of steel technologies. Time-lag and the speed of diffusion. *Technol. Forecast. Social Change 23*, 305–323.

Prigogine, I. (1995). Why irreversibility? The formulation of classical and quantum mechanics for nonintegrable systems. *Int. J. of Bifurcation and Chaos 5*(1), 3–16.

Prigogine, I. (1996). Time, chaos and the laws of nature. In P. Weingartner and G. Schurz (Eds.), *Law and prediction in the light of chaos research*, Volume LNP 473, pp. 3–9. Berlin: Springer-Verlag.

Prigogine, I. (1997). Nonlinear science and the laws of nature. *Int. J. of Bifurcation and Chaos 7*(9), 1917–1926.

Prigogine, I. and R. Lefever (1968). Symmetry breaking instabilities in dissipative systems. II. *J. Chem. Phys. 48*, 1695–1700.

Prigogine, I., R. Lefever, A. Goldbeter, and M. Herschkowitz-Kaufman (1969). Symmetry breaking instabilities in biological systems. *Nature 223*, 913–916.

Procaccia, I. (1987). *Exploring deterministic chaos via unstable periodic orbits*, Volume 2. Nucl. Phys.: proc. suppl.

Rabinovich, S., G. Berkolaiko, S. Buldyrev, A. Shehter, and S. Havlin (1997). Analytical solution of the logistic equation. *Int. J. of Bifurcation and Chaos 7*(4), 837–838.

Raha, N., J. A. Sellwood, R. A. James, and F. D. Kahn (1991). A dynamic instability of bars in disk galaxies. *Nature 352*, 411–412.

Rayleigh, L. (1883a). On maintained vibrations. *Phil. Mag. 15*, 229–235.

Rayleigh, L. (1883b). On the crispations of fluid resting upon a vibrating support. *Phil. Mag. 16*, 50–58.

Rayleigh, L. (1887). On the maintenance of vibrations by forces of double frequency, and on the propagation of waves through a medium endowed with a period structure. *Phil. Mag. 24*, 145–159.

Rayleigh, L. (1891). On the problem of random vibrations, and of random flights in one, two, or three dimensions. *Phil. Mag. 37*, 321.

Rényi, A. (1959). On the dimension and entropy of probability distributions. *Acta Math. Acad. Sci. Hung. 10*, 193–215.

Rice, S. O. (1945). Mathematical analysis of random noise. *Bell System Tech. J. 23 and 24*, 1–162.

Riess, P. (1859). Das anblasen offener rohre durch eine flamme. *Ann. Phys. Chem. 108*, 653–656.

Rijke, P. L. (1859). Notiz über eine neue art, die in einer an beiden enden offenen Röhre enthaltene luft in schwingungen zu versetzen. *Ann. Phys. Chem. 107*, 339–343.

Ritt, F. C. (1923). Permutable rational functions. *Trans. Am. Math. Soc. 25*, 399–448.

Rössler, O. E. (1976a). Chaotic behavior in simple reaction systems. *Z. Naturforsch. 31*, 259–264.

Rössler, O. E. (1976b). Chemical turbulence: Chaos in a simple reaction-diffusion system. *Z. Naturforsch. A 31*, 1168–1172.

Rössler, O. E. (1976c). Different types of chaos in two simple differential equations. *Z. Naturforsch. 31*, 1661.

Rössler, O. E. (1976d). An equation for continuous chaos. *Phys. Lett. A 57*, 397–398.

Rössler, O. E. (1979a). *Continuous chaos — four prototype equations*, Volume 316 of *Ann. N. Y. Acad. Sci.* New York: The New York Academy of Sciences.

Rössler, O. E. (1979b). An equation for hyperchaos. *Phys. Lett. A 71*(2/3), 155–159.

Rössler, O. E. (1983a). The chaotic hierarchy. *Z. Naturforsch. 389*, 788–801.

Rössler, O. E. (1983b). Macroscopic behavior in a simple chaotic Hamiltonian system. In *Dynamical systems and chaos*, Volume 179 of *Lecture notes in physics*, pp. 67–78.

Rössler, O. E. and G. C. Hartmann (1995). Attractors with flares. *Fractals 3*(2), 285–296.

Rössler, O. E., J. L. Helson, M. Klein, and C. Mira (1990). Self similar basin in continuous system. In W. Schiehlen (Ed.), *Nonlinear Dynamics in Engineering Systems*, pp. 265–273. Berlin: Springer-Verlag.

Rössler, O. E. and J. L. Hudson (1989). Self-similarity in hyperchaotic data. In E. Basar and T. H. Bullock (Eds.), *Brain Dynamics*, Volume 2 of *Springer Series in Brain Dynamics*, pp. 113–121. Berlin: Springer-Verlag.

Rössler, O. E., J. L. Hudson, and M. Klein (1989). Chaotic forcing generates wrinkled boundary. *J. Phys. Chem. 93*, 2858–2860.

Rössler, O. E., J. L. Hudson, M. Klein, and R. Wais (1988). Self-similar basin boundary in an invertible system (folded-towel map). In J. A. S. Kelso, A. J.

Mandell, and M. F. Schlesinger (Eds.), *Dynamic patterns in complex systems*, pp. 209–218. Singapore: World Scientific.

Rössler, O. E. and C. Kahlert (1979). Winfree meandering in a two-dimensional two-variable excitable medium. *Z. Naturforsch. 34*, 565–570.

Roux, J.-C. (1983). Experimental studies of bifurcations leading to chaos in the Belousov-Zhabotinsky reaction. *Physica D 7*, 57.

Rowlands, G. (1983). An approximate analytic solution of the Lorenz equations. *J. Phys. 16*, 585.

Rowlands, G. and J. C. Sprott (1992). Extraction of dynamical equations from chaotic data. *Physica D 58*, 251–259.

Ruelle, D. (1973). Some comments on chemical oscillations. *Trans. N. Y. Acad. Sci. (Ser. II) 35*(1), 66–71.

Ruelle, D. (1978). Sensitive dependence on initial condition and turbulence behavior of dynamical systems. *Ann. N. Y. Acad. Sci. 316*, 408–416.

Ruelle, D. (1979a). Ergodic theory of differentiable dynamical systems. *Publ. Phys. Math. IHES 50*, 275–306.

Ruelle, D. (1979b). Microscopic fluctuations and turbulence. *Phys. Lett. A 72*(2), 81–82.

Ruelle, D. (1981). Differentiable dynamical systems and the problem of turbulence. *Bull. Amer. Math. Soc. 5*(29-42).

Ruelle, D. (1983). Five turbulent problems. *Physica D 7*, 40.

Ruelle, D. (1985). Rotation numbers for diffeomorphisms and flows. *Ann. Inst. H. Poincaré 42*, 109–115.

Ruelle, D. (1986a). Locating resonances for axiom A dynamical systems. *J. Stat. Phys. 44*, 281.

Ruelle, D. (1986b). Resonances of chaotic dynamical systems. *Phys. Rev. Lett. 56*, 405–407.

Ruelle, D. (1987). Diagnosis of dynamical systems with fluctuating parameters. In *Proc. Roy. Soc. Ser. A*, Volume 413, Princeton, NJ, pp. 5–8.

Ruelle, D. (1989a). *Chaotic evolution and strange attractors: The statistical analysis of time series for deterministic nonlinear systems*. Cambridge: University Press.

Ruelle, D. (1989b). *Elements of the differentiable dynamics and bifurcation theory*. Boston: Academic Press.

Ruelle, D. (Ed.) (1995). *Turbulence, strange attractors, and chaos*, Volume 16 of *World Scientific Series on Nonlinear Science, Ser. A*. Singapore: World Scientific.

Ruelle, D. and F. Takens (1971). On the nature of turbulence. *Comm. Math. Phys. 20*, 167–192.

Runge, C. (1895). Über die numerische auflösung von differentialgleichungen. *Math. Ann. 46*, 167–178.

Sakaguchi, H. and K. Tomita (1987). Bifurcations of the coupled logistic map. *Prog. Theor. Phys. 78*(2), 305–315.

Sarkovskii, A. N. (1964). Coexistence of cycles of a continuous map of a line into itself. *Ukr. Mat. Z. 16*, 61–71.

Sauer, T. and J. A. Yorke (1991). Rigorous verification of trajectories for the computer simulation of dynamical systems. *Nonlinearity 4*, 961–979.

Schmelcher, P. and F. K. Diakonos (1997). Detecting unstable periodic orbits of chaotic dynamical systems. *Phys. Rev. Lett. 78*(25), 4733–4736.

Schuster and H. G. (Eds.) (1998). *Handbook of chaos control*. Wiley-VCh.

Schuster, H. G. (1988). *Deterministic chaos: An introduction*. Weinheim: VCH.

Schwartz, I. B. and I. T. Georgiou (1998). Instant chaos and hysteresis in coupled linear-nonlinear oscillators. *Phys. Lett. A 242*(6), 307–312.

Scovel, C., I. G. Kevrekidis, and B. Nicolaenco (1988). Scaling laws and the prediction of bifurcations in systems modeling pattern formation. *Phys. Lett. A 130*, 73–80.

Sellwood, J. A. (1983). Quiet starts for galaxy simulations. *J. Comp. Phy. 50*(3), 337–359.

Sellwood, J. A. (Ed.) (1989). *Dynamics of astrophysical discs*. Cambridge, UK: Cambridge University Press.

Sellwood, J. A. and A. Wilkinson (1993). Dynamics of barred galaxies. *Reports on Progress in Physics 56*(2).

Shannon, C. E. and W. Weaver (1949). *The mathematical theory of information*. Urbana, IL: University Press.

Sharif, M. N. and C. Kabir (1976). A generalized model for forecasting technological substitution. *Technol. Forecast. Social Change 8*, 353–364.

Sharkovsky, A. N. (1994). Ideal turbulence in an idealized time-delayed Chua's circuit. *Int. J. of Bifurcation and Chaos 4*(2), 303–309.

Sharkovsky, A. N. (1995). Coexistence of cycles of a contiuous map of the line into itself. *Int. J. of Bifurcation and Chaos 5*(5), 1263–1273.

Sharkovsky, A. N., P. Deregel, and L. O. Chua (1995). Dry turbulence and period-adding phenomena from a 1-D map with time delay. *Int. J. of Bifurcation and Chaos 5*(5), 1283–1302.

Shil'nikov, A. (1991). Bifurcation and chaos in the morioka-shimizu system. *Selecta Mathematica Sovietica 10*, 105–117.

Shil'nikov, A. (1993). On bifurcations of the Lorenz attractor in the shimizu-morioka model. *Physica D 62*, 338–346.

Shilnikov, A. L. and L. P. Shilnikov (1991). On the nonsymmetrical Lorenz model. *Int. J. of Bifurcation and Chaos 1*(4), 773–776.

Shilnikov, A. L., L. P. Shilnikov, and D. V. Turaev (1993). Normal forms and Lorenz attractors. *Int. J. of Bifurcation and Chaos 3*(3), 1123–1139.

Shilnikov, L. P. (1965). A case of the existence of a countable number of periodic motions. *Sov. Math. Dokl. 6*, 163–166.

Shilnikov, L. P. (1967). On the Poincaré-Birkhoff problem. *Matem. Sbornik 3*, 353–371.

Shilnikov, L. P. (1984). *Bifurcation theory and turbulence*, Volume 2. Harward Academic Publishers.

Sirovich, L. and P. K. Newton (1986). Periodic solutions of the Ginzburg-Landau equation. *Physica D 21*, 115.

Skiadas, C. H. (1985). Two generalized rational models for forecasting innovation diffusion. *Technol. Forecast. Social Change 27*, 39–61.

Skiadas, C. H. (1986). Innovation diffusion models expressing asymmetry and/or positively or negatively influencing forces. *Technol. Forecast. Social Change 30*, 313–330.

Skiadas, C. H. (1987). Two simple models for early and middle stage prediction of innovation diffusion. *IEEE Trans. Eng. Manag. 34*, 79–84.

Skiadas, C. H. (1994). Methods of growth functions formulation. In R. Gutiérrez and M. Valderrama (Eds.), *Selected topics on stochastic modelling*. Singapore: World Scientific.

Skiadas, C. H. (1995). A Lagrangian approach for the selection of growth functions in forecasting. In J. Janssen, C. H. Skiadas, and C. Zopounidis (Eds.), *Advances in stochastic modelling and data analysis*. Dordrecht: Kluwer Acad. Publishers.

Skiadas, C. H. (2007). *Recent advances in stochastic modeling and data analysis*, Chapter Exploring and simulating chaotic advection: A difference equations approach. Singapore: World Scientific.

Skiadas, C. H. and A. N. Giovanis (1997). A stochastic Bass innovation diffusion model for studying the electricity consumption in Greece. *Appl. Stoch. Mod. Dat. Analys. 13*, 85–101.

Skiadas, C. H., A. N. Giovanis, and J. Dimoticalis (1993). A sigmoid stochastic growth model derived from the revised exponential. In J. Janssen and C. H. Skiadas (Eds.), *Applied stochastic models and data analysis*, Volume II. Singapore: World Scientific.

Skiadas, C. H., A. N. Giovanis, and J. Dimoticalis (1994). Investigation of stochastic differential models: the Gompertzian case. In R. Gutiérrez and M. Valderrama (Eds.), *Selected topics on stochastic modelling*. Singapore: World Scientific.

Skiadas, C. H., C. Zopounidis, and J. Dimoticalis (1993). Exploring the fitting and forecasting performance of sigmoid forecasting models: A comparative study in the Greek economic data series. In J. Janssen and C. H. Skiadas (Eds.), *Applied stochastic models and data analysis*, Volume II. Singapore: World Scientific.

Smale, S. (1967). Differentiable dynamical systems. *Bull. Am. Math. Soc. 73*, 747–817.

Smale, S. (1976). On the differential equations of species in competition. *J. Mathem. Biology 3*, 5–7.

Smale, S. (1977). Dynamical systems and turbulence. In Chorin, Marsden, and Smale (Eds.), *Turbulence seminar*, Volume 615, pp. 71–82. Berlin: Springer-Verlag.

Smale, S. (1980). *The mathematics of time; Essays on dynamical systems, economic processes, and related topics*. New York: Springer-Verlag.

Smale, S. and R. F. Williams (1976). The qualitative analysis of a difference equation of population growth. *J. Mathem. Biology 3*, 1–4.

Sondhauss, C. (1850). Über die schallschwingungen der luft in erhitzten glasröhren und in gedeckten pfeifen von ungleicher weite. *A. Phys. Chem. 79*, 1–34.

Sonis, M. (1996). Once more on Hénon map: Analysis of bifurcations. *Chaos Solitons Fractals 7*(12), 2215–2234.

Sophianopoulos, D. S., A. N. Kounadis, and A. F. Vakakis (2002). Complex dynamics of perfect discrete systems under partial follower forces. *Int. J. Non-Linear Mech. 37*(7), 1121–1138.

Spano, M. L. and W. L. Ditto (1994). Controlling chaos. *AIP Conf. Proc. 296*, 137–156.

Sparke and J. S. Gallagher (2000). *Galaxies in the universe*. Cambridge, UK: Cambridge University Press.

Sprott, J. C. (1994a). Predicting the dimension of strange attractors. *Phys. Lett. A 192*(5-6), 355–360.

Sprott, J. C. (1994b). Some simple chaotic flows. *Phys. Rev. E 50*(2), 647–650.

Sprott, J. C. (1997). Simplest dissipative chaotic flow. *Phys. Lett. A 228*, 271–274.

Stewart, I. (1989). *Does God play dice? The mathematics of chaos*. Oxford: Blackwell.

Strogatz, S. H. (1994). *Nonlinear dynamics and chaos*. Redwood City: Addison-Wesley.

Swinney, H. L. (1978). Hydrodynamic instabilities and the transition to turbulence. *Theor. Phys. Supp. 64*, 164.

Swinney, H. L. and J. P. Gollub (1978). Transition to turbulence. *Phys. Today 31*, 41.

Tata, F. and J. C. Vassilicos (1991). Is there chaos in economic time series? A study of the stock and foreign exchange markets. discussion paper 120, Financial markets group, London School of Economics.

Taylor, M. A. and I. G. Kevrekidis (1991). Some common dynamic features of coupled reacting systems. *Physica D 51*, 274–292.

Thompson, J. and H. Stewart (1987). *Nonlinear dynamics and chaos*. New York: Wiley.

Thompson, J. M. T. and S. R. Bishop (Eds.) (1994). *Nonlinearity and chaos in engineering dynamics*. Chichester: Wiley.

Tjahjadi, M. and J. M. Ottino (1991). Stretching and breakup of droplets in chaotic flows. *J. Fluid Mech. 232*, 191–219.

Tomita, K. and H. Daido (1980). Possibiliy of chaotic behaviour and multi-basins in forced glycolytic oscillations. *Phys. Lett. A 79*, 133.

Tomita, K. and D. K. (1979). Thermal fluctuation of a self-oscillating reaction system entrained by a periodic external force. *Prog. Theor. Phys. 61*, 825.

Toomre, A. (1963). On the distribution of matter within high flattened galaxies. *Astrophys. J. 138*, 385–392.

Toomre, A. (1964). On the gravitational stability of a disk of stars. *Astrophys. J. 139*, 1217–1238.

Tsiganis, K. and H. Varvoglis (2000). Chaotic evolution of (719) Albert, the recently recovered minor planet. *Astron. Astrophys. 361*(2), 766–769.

Tsonis, A. A. (1988). On the dimension of the weather attractor. (Special issue). *Ann. Geophys.*, 208.

Tsonis, A. A. (1992). *Chaos: From theory to applications*. New York: Plenum.

Tsonis, A. A. (1996). Dynamical systems as models for physical processes. *Complexity 1*(5), 23–33.

Tsonis, A. A. (2001a). The impact of nonlinear dynamics in the atmospheric sciences. *Int. J. Bifurcation Chaos 11*(4), 881–902.

Tsonis, A. A. (2001b). Probing the linearity and nonlinearity in the transitions of the atmospheric circulation. *Nonlinear Process Geophys. 8*(6), 341–345.

Tsonis, A. A. and J. B. Elsner (1988). The weather attractor on very short time scales. *Nature 333*, 545–547.

Tsonis, A. A. and J. B. Elsner (1989). Chaos, strange attractors, and weather. *Bull. Amer. Meteor. Soc. 70*, 14.

Tsonis, A. A. and J. B. Elsner (1990). Multiple attractors, fractal basins and longterm climate dynamics. *Contrib. Atmos. Phys. 63*, 171.

Tsonis, A. A. and J. B. Elsner (1992a). Estimating the dimension of weather and climate attractors: Important issues about the procedure and interpretation. *J. Atmos. Sci. 50*, 2549–2555.

Tsonis, A. A. and J. B. Elsner (1992b). Nonlinear prediction as a way of distinguishing chaos from random fractal sequences. *Nature 359*, 217–220.

Tucker, W. (1999). The Lorenz attractor exists. *C. R. Acad. Sci. Ser. I Math. 328*(12), 1197–1202.

Turing, A. M. (1952). The chemical basis of morphogenesis. *Phil. Trans. Roy. Soc. 237*, 5.

Tyson, J. J. (1984). Relaxation oscillations in the revised oregonator. *Journ. Chem. Phys. 80*, 6079–6082.

Tzafestas, S. and E. Tzafestas (2001). Computational intelligence techniques for short-term electric load forecasting. *J. Intell. Robot. Syst. 31*(1-3), 7–68.

Ueda, Y. (1979). Randomly transitional phenomena in the system governed by Duffing's equation. *J. Stat. Phys. 20*, 181–190.

Ueda, Y. (1980). *Explosion of strange attractors exhibited by Duffing's equation*, Volume 357 of *Ann. N. Y. Acad. Sci.* New York: The New York Academy of Sciences.

Ueda, Y. (1985). Random phenomena resulting from nonlinearity in the system described by Duffing's equation. *Int. J. Nonlinear Mech. 20*(5-6), 481–491.

Ueda, Y. (1992a). *The road to chaos*. Santa Cruz, CA: Aerial Press.

Ueda, Y. (1992b). Strange attractors and the origin of chaos. *Nonlinear Science Today 2*, 1–16.

Ueda, Y., C. Hayashi, and N. Akamatsu (1973). Computer simulation of nonlinear ordinary differential equations and nonperiodic oscillations. *Electronics and Commun. in Japan 56*, 27.

Ueda, Y., H. Nakajima, T. Hikihara, and H. B. Stewart (1988). Forced two-well potential Duffing's oscillator. In F. M. A. Salam and M. L. Levi (Eds.), *Dynamical systems approaches to nonlinear problems in systems and circuits*, pp. 128–137. Philadelphia, PA.: Soc. for Industrial and Appl. Math. (SIAM).

Ueda, Y., H. Ohta, and H. B. Stewart (1994). Bifurcation in a system described by a nonlinear differential equation with delay. *Chaos 4*(1), 75–83.

Uhlenbeck, G. E. and L. S. Ornstein (1930). On the theory of the Brownian motion. *Phys. Rev. 36*(3), 823–841.

Ulam, S. M. and J. von Neumann (1947). On combinations of stochastic and deterministic processes. *Bull. Am. Math. Soc. 53*, 1120.

Ushiki, S. (1982). Central difference scheme and chaos. *Physica D 4*, 407.

Ushiki, S. and R. Lozi (1987). Confinor and anti-confinor in constrained "Lorenz" system. *Japan. J. Appl. Math. 4*, 433–454.

Utida, S. (1957). Population fluctuation, an experiment and theoretical approach. *Cold Spring Harb. Symp. Quant. Biol. 22*, 139.

van der Pol, B. (1927). Forced oscillations in a circuit with non-linear resistance. *Phil. Mag. Ser. 7 3*, 65–80.

van der Pol, B. and J. van der Mark (1927). Frequency demultiplication. *Nature 120*, 363–364.

Vayenas, D. V. and S. Pavlou (1999). Chaotic dynamics of a food web in a chemostat. *Math. Biosci. 162*(1-2), 69–84.

Vayenas, D. V. and S. Pavlou (2001). Chaotic dynamics of a microbial system of coupled food chains. *Ecol. Model. 136*(2-3), 285–295.

Verhulst, F. (1990). *Nonlinear differential equations and dynamical systems*. New York: Springer-Verlag.

Verhulst, P. F. (1845). Recherches mathématiques sur la loi d'acroissement de la population. *Mem. Acad. Roy. Belg. 18*.

Voglis, N. and G. J. Contopoulos (1994). Invariant spectra of orbits in dynamical systems. *J. Phys. A 27*, 4899–4909.

Volterra, V. (1959). *Theory of functionals and of integral and integro-differential equations*. New York: Dover.

Voyatzis, G. and S. Ichtiaroglou (1999). Degenrate bifurcations of resonant tori in Hamiltonian systems. *Int. J. of Bifurcation and Chaos 9*(5), 849–863.

Vozikis, C. L. (2001). The transition phase of the deviation vector of nearby orbits. *J. Phys. A-Math. Gen. 34*(7), 1513–1527.

Vrahatis, M. N. (1995). An efficient method for locating and computing periodic orbits of nonlinear mappings. *J. Comput. Phys. 119*(1), 105–119.

Vrahatis, M. N., T. C. Bountis, and M. Kollmann (1996). Periodic orbits and invariant surfaces of 4D nonlinear mappings. *Int. J. of Bifurcation and Chaos* 6(8), 1425–1437.

Walker, G. (1931). On periodicity in series of related terms. In *Proc. Roy. Soc. London A*, Volume 131, pp. 518–532.

Walker, G. and E. W. Bliss (1937). World weather IV. *Mem. Roy. Met. Soc. 4*, 119–139.

Wang, M. C. and G. E. Uhlenbeck (1945). On the theory of the Brownian motion II. *Rev. of Mod. Phys. 17*(2and3), 323–342.

Wegmann, K. and O. E. Rössler (1978). Different kinds of chaotic oscillations in the Belousov-Zhabotinskii reaction. *Z. Naturforsch. 33*, 1179.

Weierstrass, F. (1872). Über kontinuierliche funktionen eines reellen arguments, die für keinen wert des letzteren einen bestimmten differential quotienten besitzen. In *Mathematische Werke II*, pp. 71–74.

West, R. W. (1924). The action between bromine and malonic acid in aqueous solution. *J. Chem. Soc. 125*, 1277–1282.

White, R. B., S. Benkadda, S. Kassibrakis, and G. M. Zaslavsky (1998). Near threshold anomalous transport in the standard map. *Chaos 8*, 757–767.

Wiener, N. (1930). Generalized harmonic analysis. *Acta Math. 55*, 117–258.

Wiener, N. (1938). The homogeneous chaos. *Amer. J. Math. 60*, 897.

Wiener, N. (1949). *The extrapolation, interpolation and smoothing of stationary time series*. New York: Wiley.

Wiener, N. (1958). *Nonlinear problems in random theory*. New York: Wiley.

Williams, R. F. (1979a). *The bifurcation space of the Lorenz attractor*, Volume 316 of *Ann. N. Y. Acad. Sci.* New York: The New York Academy of Sciences.

Williams, R. F. (1979b). The structure of Lorenz attractors. *Publ. Math. IHES 50*, 321.

Winfree, A. T. (1972). Spiral waves of chemical activity. *Science 175*, 634–636.

Winfree, A. T. (1973). Scroll-shaped waves of chemical activity in three dimensions. *Science 181*, 937–939.

Winfree, A. T. (1974a). Rotating chemical reaction. *Sci. Amer. 230*, 82–95.

Winfree, A. T. (1974b). Rotating solutions to reaction/diffusion equations in simply-connected media. In D. S. Cohen (Ed.), *Mathematical aspects of chemical and biochemical problems and quantum chemistry*, Volume 8, pp. 13–31. SIAM - AMS.

Winfree, A. T. (1978). Stable rotating patterns of reaction and diffusion. *Theoret.*

Chem. 4, 1–51.

Wisdom, J. (1987a). Chaotic behavior in the solar system. In M. V. Berry, I. C. Percival, and N. O. Weiss (Eds.), *Dynamical Chaos, Proc. of a Roy. Soc. Disc. Meeting, Feb. 1987*, pp. 109–129. Princeton, NJ: University Press.

Wisdom, J. (1987b). Chaotic dynamics in the solar system. *Icarus 72*, 241.

Yamada, T. and Y. Kuramoto (1976). A reduced model showing chemical turbulence. *Prog. Theor. Phys. 56*, 681.

Yamaguti, M. and S. Ushiki (1981). Chaos in numerical analyses of ordinary differential equations. *Physica D 3*, 618–626.

Yannacopoulos, A. N., J. Brindley, J. H. Merkin, and M. J. Pilling (1996). Approximation of attractors using Chebyshev polynomials. *Physica D 99*(2-3), 162–174.

Yannacopoulos, A. N., I. Mezic, G. Rowlands, and G. P. King (1998). Eulerian diagnosis for Lagrangian chaos in three-dimensional Navier-Stokes flows. *Phys. Rev. E 57*(1), 482–490.

Yorke, J. A., E. D. Yorke, and J. Mallet-Paret (1987). Lorenz-like chaos in a partial differential equation for a heated fluid loop. *Physica D 24*, 279.

Yoshida, H., A. Ramani, and B. Grammaticos (1987). Painlevéresonances vs. Kowalevski exponents: exact results on singularity structure and integrability of dynamical systems. *Acta Appl. Math. 8*, 75.

Yuan, G. C. and J. A. Yorke (2000). Collapsing of chaos in one dimensional maps. *Physica D 136*(1-2), 18–30.

Zaslavsky, G. M. (1978). The simplest case of a strange attractor. *Phys. Lett. A 69*, 145–147.

Zaslavsky, G. M. (1993). Self-similar transport in complete chaos. *Phys. Rev. E 48*, 1683.

Zaslavsky, G. M. (1999). Chaotic dynamics and the origin of statistical laws. *Physics Today 52*(8), 39–45.

Zaslavsky, G. M. and B. V. Chirikov (1972). Stochastic instabillities in nonlinear oscillations. *Sov. Phys. Usp. 14*, 549.

Zhabotinskii, A. M. (1964). Periodic course of oxidation of malonic acid in solution (investigation of the kinetics of the reaction of Belousov). *Biophysics 9*, 329–335.

Zhabotinskii, A. M., A. N. Zaikin, M. D. Korzukhin, and G. P. Kreitser (1971). Mathematical model of a self-oscillating chemical reaction (oxidation of bromomalonic acid with bromate catalyzed by cerium ions). *Kinetics and Catalysis 12*, 516–521.

Index

T - #0369 - 071024 - C368 - 234/156/16 - PB - 9780367386658 - Gloss Lamination